108th Annual Meeting of the South Dakota Academy of Science 2023

Volume 102

Vermillion, South Dakota, USA
23 October 2023

Editors:

Robert Tatina
Michael Barnes
Dave Bergmann
Steve Chipps
Andy Detwiler
Andy Farke
Donna Hazelwood
Tim Mullican
Jeffrey Palmer
Fedora Sutton

ISBN: 979-8-3313-0003-6

Printed from e-media with permission by:

Curran Associates, Inc.
57 Morehouse Lane
Red Hook, NY 12571

Some format issues inherent in the e-media version may also appear in this print version.

Printed with permission by Curran Associates, Inc. (2025)

For permission requests, please contact South Dakota Academy of Science
at the address below.

South Dakota Academy of Science
308 Orchard Blvd.
Sawyer MI 49125

Phone: (605) 996 7670

southdakotaacademyofscience.wildapricot.org/

Additional copies of this publication are available from:

Curran Associates, Inc.
57 Morehouse Lane
Red Hook, NY 12571 USA
Phone: 845-758-0400
Fax: 845-758-2633
Email: curran@proceedings.com
Web: www.proceedings.com

Proceedings
of the
South Dakota Academy of Science

Volume 102
2023

Published by the South Dakota Academy of Science
Academy Founded 22 November, 1915
Academy Website: ww.sdaos.org

Graphic Designer
Tom Holmlund

Printer
Minuteman Press, Sioux Falls, SD

ISSN 0096-378X (print)
ISSN 0096-1947 (online)

Copies of articles are available at EBSCO host and at www.sdaos.org

The Proceedings of the South Dakota Academy of Science

Since its beginning in 1915, the South Dakota Academy of Science has been a forum for the promotion of scientific teaching, research, and service in the state of South Dakota. The South Dakota Academy of Science meets each year for the purpose of conducting annual business, presenting awards, and facilitating the interchange of ideas, information, and results from scientists and students of science. The minutes of the annual meeting and the abstracts and full papers of presentations are published annually in the Proceedings of the South Dakota Academy of Science, which is uploaded to the Academy website (www.sdaos.org). Although presentation at the annual meeting is strongly encouraged, each year a limited number of full papers may be considered for publication that have not been presented. The Instructions for Authors can be viewed at www.sdaos.org.

This volume contains the minutes of the annual meeting which was held at the University of South Dakota October 29, 2023, and various reports. Additionally, it contains the full papers and abstracts of talks and posters presented at the annual meeting.

TABLE OF CONTENTS

Full Papers Submitted for Presentation at the 108th Annual Meeting of the South Dakota Academy of Science

Abstracts of Research Papers and Posters Submitted for Presentation at the 108th Annual Meeting of the South Dakota Academy of Science

COMBINED MINUTES
108TH ANNUAL MEETING
SOUTH DAKOTA ACADEMY OF SCIENCE
23 OCTOBER 2023
UNIVERSITY OF SOUTH DAKOTA
VERMILLION, SD

A. Opening
   1. Welcome by President Tim Mullican. Tim welcomed everyone to the 108th SDAS Annual Business meeting hosted by USD on the USD campus in the MUC Ballroom C, October 23, 2023 at 3:00 pm CT. He praised the Local Arrangements Co Chairs, Dave Swanson and Bruce Cuevas for the best SDAS meeting since before COVID noting the Missouri River Science and Management Symposium, the Keynote Speaker, Tyler Kelley, for his presentation "*Holding back the River.*", and the oral papers and posters. He encouraged everyone to take care as we literally weather the (snow) storm occurring outdoors. He shared his enthusiasm to build on the successes of this meeting and move the SDAS forward.
   2. Passing of the gavel to Incoming President Tom Montoya. Tim passed the gavel to Incoming President Tom Montoya. Tom by reciprocated sharing a plaque honoring Tim for his tenure as President during a challenging time through uncharted territory...
   3. Welcome by Incoming President Tom Montoya. Tom Montoya welcomed everyone to the meeting and suggested that in light of the snowstorm out of doors, that the reports which were presented last at the Executive Council Meeting be skipped. Hearing no concerns, he moved on to approval of the agenda.

B. Approval of agenda
   1. It was moved and seconded that the agenda be approved. The motion carried.

C. Reports
   1. Secretary's report by Donna Hazelwood
      a. The SDAS Annual Meeting was hosted by USD on 27 and 28 Oct. 2023. Copies of individual certificates for the annual Poster and Paper Competitions were provided to the treasurer. The 2023 annual South Dakota Not-for-Profit report and filing fee were forwarded to the South Dakota Secretary of State. The 2023 SDAS Annual Meeting Executive Council meeting will be 4:00 pm on Friday 27 October 2023. The date for the next upcoming SDAS Executive Council meeting is TBA. Respectfully submitted, Donna Hazelwood, SDAS secretary
   2. Treasurer's report by Jeffrey Palmer
   3. Proceedings Editor's report with revisions by Robert Tatin

4. Fellows Nomination Committee report by Bob Tatina
5. No nominations received for SDAS Fellow for 2023.
6. Membership Committee report by Tim Mullican.
7. Social media report by Tim Mullican
8. Publicity report by Dave Bergman
9. Web Master's report by Tim Mullican
10. Audit Committee report by Mark Gabel.
11. Resolutions Committee report by Donna Hazelwood

D. Old/Continuing Business
   1. The SDAS recognizes and congratulates the undergraduate and graduate student awardees from the 2022 SDAS Annual Meeting as follows. We congratulate the awardees of Poster Judging 2022
      a. The first-place winner of the poster competition was Julienne Korst (UG) (SDSU) for: FATE OF EASTERN REDCEDAR SEEDS. Lan Xu, Julienne Korst (UG), and Colin Sachen (UG) (SDSU)
      b. The second-place winner of the poster competition was Jack Rabern (UG) (SDSU) for: BUMBLE BEE (*BOMBUS* LATREILLE) SPECIES DIVERSITY IN MATURE AND RECENTLY-ESTABLISHED NATIVE PLANTINGS IN BROOKINGS, SD. Jack Rabern (UG), Max Miller (UG), and Abigail P. Martens (SDSU)
   2. We congratulate the awardees of Oral Paper Judging 2022;
      a. First Place winner of the Graduate Student Competition Shuyi He (USD) for: SEMICONDUCTING POLYMER DOTS L-LACTATE TRANSDUCER BY ENZYMATIC CASCADE REACTION SYSTEM. Shuyi He and Steven Wu (USD)
      b. First place winner of the Undergraduate Student Oral Paper Oleksandra Rachyaska (DSU) for:
      C. PATHOGENICITY, DETECTION AND PREVENTION OF PYTHIUM SEED ROT IN ALFALFA. Oleksandra Rachynska (UG), Travis Rebstock , Conner Tordsen, Jennifer Giles, and Andrew Sathoff (DS)DSU

E. New Business
   1. Jeff Palmer moved and Paula Mazzer seconded a motion to increase page charges for the SDAS Proceedings from $20.00 per page to $25.00 per page beginning with the 2024 Proceedings. Tim Mullican moved and Bruce Cuevas seconded approval of this motion. The motion carried.
   2. Poster and paper presentation awards
      a. Paula Mazer moved and Bruce Cuevas seconded to award first, second and third place awards to both undergraduate and graduate student poster presenters and to award similar awards to a combined undergraduate and grad student pool of paper presenters. The motion carried.
         i. Awards for first, second ands third place are $300.00, $200.00, and $100.00 respectively.
      b. Paul Johnson moved and Paula Mazzer seconded approval of the motion. The motion passed.

3. SDAS New Researcher Grant Recipient
   a. We congratulate Reza Mehrabian. The title of his project is "Developing an Invariant Extended Kalman Filter Estimation System for Three Distinct Robot Variants." Reza is a new faculty member at SDSMT.
4. SDAS Elections
   a. Member-at-Large 2023-2025
      i. Continuing Mark Gabel, BHSU
      ii. Continuing, Adam Sturlaugson, USF
      iii. New, George Nora, NSU
      iv. New, Eric Pulis, NSU
      v. New, Abagail Martens, SDSU
      vi. Open four positions
      vii. Open graduate student Member-at-Large (two positions one or two year)
   b. Open positions
      i. Two Co-chair Membership Committee
      ii. Graduate student Co-chair Social Media committee (could accommodate two)
   c. Ballot for SDAS Executive Council Elections, Slate of Nominees
   d. Member-at-Large 2023-2025
      i. Continuing Mark Gabel, BHSU
      ii. Continuing, Abagail Martens, SDSU
      iii. Continuing, Adam Sturlaugson, USF
      iv. New, George Nora, NSU
      v. New, Eric Pulis, NSU
   e. Jeff palmer moved and Bruce Cuevas seconded approval of the slate of nominees for Member-at-Large 2023-2025. The motion carried.
5. Poster awards
   a. Undergraduate student poster awards
      i. First place: CHARACTERIZATION OF FLUORESCENT PROTEIN-EXPRESSING INFLUENZA VIRUSES. **Caycee Schneider (UG)** and Victor Huber. Chemistry Department, Mount Marty University, Yankton, SD.
      ii. Second place: DETERMINATION OF MINIMUM INHIBITORY CONCENTRATION (MIC) AND MINIMUM BACTERICIDAL CONCENTRATION (MBC) OF *SALVIA ROSMARINUS* USING A 96 WELL PLATE ASSAY. **Elisabeth McGill (UG)** and Kathleen A. Gibson. Mount Marty University, Yankton, SD.
      iii. Third place: CYTOTOXICITY AND DETERMINATION OF THE MINIMUM BACTERICIDAL CONCENTRATION (MBC) FOR *COLEUS SCUTELLARIOIDES*. **Lexa Burtzlaff (UG)** and Kathleen A. Gibson, Mount Marty University, Yankton, SD.

   b. Graduate student poster awards
      i. First place: DO WE NEED A SALINE BOLUS? THE NECESSARY REFINEMENT OF PEDIATRIC PROPOFOL SEDATION Laura Nelson, **Madigan Moore,** Andrew Guymon and Mir Ali. Sanford School of Medicine, University of South Dakota, Vermillion, SD.
      ii. Second place: EXPLORING THE ROLE OF BACTERIOPHAGE IN THE RGG1-MEDIATED GENE REGULATION OF *STREPTOCOCCUS PYOGENES*. **Yuan Tian,** Andrea Herrera and Michael Chausse. Sanford School of Medicine. University of South Dakota, Vermillion, SD.
      iii. Third place: FIELD-SCALE DETECTION OF MULTIPLE AGRICULTURAL MANAGEMENT PRACTICES THROUGH MACHINE LEARNING ALGORITHMS IN EASTERN SOUTH DAKOTA. **Khushboo Jain,** Ranjeet John, Nathan Torbick, Venkatesh Kolluru, Sakshi Saraf, Geoff Henebry, Abhinav Chandel, and Meghann Jarchow. Department of Sustainability & Environment, University of South Dakota, Vermillion, SD.
   c. Oral Contributed Paper Awards
      i. First place: VARIATION IN METABOLIC RATES AND BODY AND TISSUE MASSES IN RESPONSE TO FLUCTUATING COLD IN HOUSE SPARROWS. **Chelsi Marolf** and David Swanson. Department of Biology, University of South Dakota, Vermillion, SD.
      ii. Second Place: BMP 9/ALK1 SIGNALING IN HUMAN HIGHLYPROLIFERATIVE CELLS. **Oleksandra Rachynska (UG),** Calvin Vary and Michayla Moore. Dakota State University, Madison, SD
      iii. Third Place: AK072019: A MOLECULAR INHIBITOR OF MRSA ACETATE KINASE EXHIBITING CELLULAR-LEVEL EFFICACY AGAINST STAPHYLOCOCCUS AUREUS, **Madison Kovar (UG),** Mitchell Lonneman (UG), and Chun Wu. Mount Marty University, Yankton, SD.
6. Resolutions: Committee
   a. The Members of the Academy sincerely and gratefully thank the following:1). Local Planning and Organizing Committee Co-Chairs Bruce Cuevas and Dave Swanson, who through dedicated teamwork hosted a well-planned, well organized and executed SDAS Annual meeting on the USD campus, while also making available Zoom connections where feasible. The SDAS is sincerely grateful. 2). Sincere special heartfelt thanks to long-standing well organized and timely Proceedings Editor Bob Tatina for his continued excellence and dedication organizing and compiling the schedule for papers and posters and for his outstanding dedication as editor of the Proceedings of the SDAS. The SDAS also thanks Bob for chairing the

Fellows Nomination Committee. 3) Sincere gratitude and recognition to Past President Tim Mullican (2021- 2023) for his tireless dedicated active and pioneering service to the SDAS through changing times, and for fostering community and science throughout the year. Always going above and beyond. In addition, sincere thanks to Tim for also serving as competent and dedicated Webmaster, Chair of Membership and chair of Social Media committees. 4) We thank outgoing Past-President Vicki Geiser for her years of dedicated service to the SDAS. 5) special thanks to Donna Hazelwood for continuing to serve as Secretary; 6) special thanks to the dedicated, well organized and meticulous Jeff Palmer, Treasurer; 7) thanks to Tim Mullican for doing dedicated double duty and continuing to serve as Web Master; 8) we thank Fellows Committee Chair Bob Tatina for service to the SDAS Fellows and membership and for inviting SDAS Fellows to nominate members; 9) thanks to all members of the SDAS Executive Council for ongoing generous service to the SDAS, 10) thanks to all members of the SDAS for ongoing or new support of the SDAS, 11) we thank the Resolution Committee: Donna Hazelwood; 12) we thank Mark Gabel and Donna Hazelwood for service on the Audit Committee, 13) we thank members of the Poster Judging Committee, Bruce Cuevas and Paula Mazzer; 14) finally, we congratulate the awardees of Poster Judging 2023.

7. Upcoming SDAS Annual meetings in person and Zoom format
    a. Upcoming SDAS Annual Meetings
        i. 109th Annual Meeting in-person/Zoom hosted by NSU 19-20 April 2024
        ii. Contacts: Local Arrangements co-chairs George Nora and Eric Pulis
        iii. 110th Annual Meeting hosted by SDSM&T 2025
        iv. 111th Annual Meeting hosted by USF 2026
        v. 112th Annual Meeting hosted by DWU 2027
        vi. 113th Annual Meeting hosted by SDSU 2028
        vii. 114th Annual Meeting hosted by MMU 2029
        viii. 115th Annual Meeting hosted by SDAS EC Chamberlain
    b. SDAS 2024 Executive Council Winter Meeting TBA 2024
        i. Format Zoom

F. Adjourn meeting
1. Bruce Cuevas moved and Paul Johnson seconded that the meeting be adjourned. The motion carried.

Respectfully submitted,
Donna Hazelwood
SDAS Secretary

TREASURER'S REPORT

01 Jan 2023 - 31 Dec 2023

01-Jan-23	Checking Account Balance	$10,494.38
	Certificate of Deposit	$9,178.69
	Petty Cash	$0.00
	Total Assets	$19,673.07

	Income	Expense	Date	Check #	Payee
Plaques for Fellows		166.47	08-Feb-23	1473	Tim Mullican
Eastern SD Science & Engineering Fair		240.00	21-Mar-23	1482	Science Fair Awards
Northern Science Fair		240.00	30-Mar-23	1474-81	Science Fair Awards
Domestic Nonprofit Filing Fee		10.00	27-May-23	1205	SD Secretary of State
New Researcher Grant - Donation	3000.00		16-Oct-23		
Fall Meeting - Room Opening Fee		52.00	24-Oct-23	1483	USD
Fall Meeting - Food		1,390.77	24-Oct-23	1484	USD - Catering
Carbon Copy Registration Forms		3 0.00	26-Oct-23	1485	DSU
Keynote Speaker Donation	3,200.00		01-Nov-23	1486	USD
New Researcher Grant Award		3,000.00	02-Nov-23	1487	SDSMT
Plaque for Past President		79.65	02-Nov-23	1488	Thomas Montoya
Meeting Registrations and Dues		35.00	06-Nov-23		
Transfer from PayPal	2000.00		06-Nov-23		
Deposit Slips		94.26	08-Nov-23		Harland Clarke CHK
Meeting Registrations / Keynote	1800.00		13-Nov-23		
Oral Paper - 3rd Place		100.00	14-Nov-23	1489	Madison Kovar
Oral Paper - 2nd Place		200.00	14-Nov-23	1490	Oleksandra Rachynska
Oral Paper - 1st Place		300.00	14-Nov-23	1491	Chelsi Marolf
Grad Poster - 3rd		100.00	16-Nov-23	1492	Khushboo Jain
Grad Poster - 2nd		200.00	16-Nov-23	1493	Yuan Tian
Grad Poster - 1st		300.00	16-Nov-23	1494	Madigan Moore

Undergrad Poster - 3rd		100.00	17-Nov-23	1495	Lexa Burtzlaff
Undergrad Poster - 2nd		200.00	17-Nov-23	1496	Elisabeth McGill
Undergrad Poster - 1st		300.00	17-Nov-23	1497	Caycee Schneider
Totals	**7635.00**	**10303.15**			

31-Dec-23	Checking Account Balance	$7,826.23
	Certificate of Deposit	$9,178.69
	Petty Cash	$0.00
	Total Assets	$17,004.92

Respectfully Submitted,
Jeffrey S. Palmer, Treasurer

PROCEEDINGS EDITOR'S REPORT 2023

1 Jan 2023 to 21 Oct 2023

A. The following libraries were mailed a complimentary paper copy of the 2022 Proceedings: AC, BHSU, DSU, DWU, MMC, NSU, OLC, USF, NSU, SBC, SGU, SDSMT, SDSU (2 copies), SWC, UND, USD (2 copies), SD State Library, Am. Mus. Nat. Hist., U of WA and the Booth Fish Hatchery.
B. The following indexing/ abstracting services were mailed complimentary paper copies of the 2021 Proceedings: EBSCOhost, SAGE Publishing (Abstracts in Anthropology), Thompson Scientific (Zoological Record)
C. Electronic files of the 2022 Proceedings have been published to the SDAoS website.
D. For the 2022Proceedings:
   1. No. of paper copies printed = 26 (2021: 30)
   2. No. of pages = 74 (2020: 140)
   3. No. of full papers published = 2 (2020: 8)
   4. No. of abstracts published = 9 (2020: 14)
   5. No. of titles only published = 0 (2020: 2)
   6. No. paper copies distributed to libraries, abstracting services, individuals, etc.= 26
   7. 50 pages invoiced @ $20.00 per page = $1000.00 (Total paid = $1000.00)
E. First call for papers sent July 1; second call sent out in August 5; final call September 2.

F. 2023 Schedule of Events sent October to entire mailing list. 100 paper copies printed by MinuteMan Press, Sioux Falls and sent to USD.
G. Proceedings Ledger

	1/1/2023	**Beginning Balance**		**6302.90**
	Date		**Expenses**	**Income**
DEP	5/15/2023	Higgins & Gabriel page charges		440.00
DEP	5/31/2023	Page charges—6 checks		120.00
2658	6/4/2023	Robert Tatina—Dropbox subscription	119.88	
2659	6/13/2023	Tom Holmund—Layout & design 2022 Proc	585.00	
2660	7/28/2023	Minuteman Press—print & mail 2022 PSAS	709.76	
DEP	8/8/2023	Page charges-		440.00
2661	10/20/2023	Miner Solutions—website hosting	600.00	
2662	10/20/2023	Minute Man Press—annual meeting program	105.83	
2663	10/31/2023	SDAS—speaker fee	1200.00	
		Total	**3320.47**	**1000.00**
	12/31/2023	**Ending Balance**		**3982.43**

Submitted by Robert Tatina, Editor

2023-2025 EXECUTIVE COUNCIL

President 2023-2025	Thomas Montoya, SDSM&T, Electrical Engineering 394-2459, Thomas.Montoya@sdsmt.edu
Vice President 2023-2025	Paula Mazzer. DWU, Chemistry. 995-2899, Paula.Mazzer@dwu.edu
Past President 2023-2025	Tim Mullican, DWU, biology 995-2715, Tim.Mullican@dwu.edu
Secretary	Donna Hazelwood, DSU, Biology, (retired) 256-5187, donna.hazelwood@dsu.edu
Treasurer	Jeffrey Palmer, DSU, Mathematics 256-5190, Jeff.Palmer@dsu.edu
Proceedings Editor	Robert Tatina, DWU (retired) 269-426-3744 Robert.Tatina@dwu.edu
Chair Fellows Nominations Committee	Robert Tatina, DWU (retired) 269-426-3744 Robert.Tatina@dwu.edu
Co-Chair Membership Committee	Open
Co-Chair Membership Committee	Open
Co-Chair Social Media Committee	Tim Mullican, DWU, biology 995-2715, Tim.Mullican@dwu.edu
Co-Chair Social Media Committee	Grad student open
Publicity	Dave Bergmann, BHSU, Biology, 642-6465 Dave.Bergmann@bhsu.edu
Publicity	Robert Tatina, DWU biology (retired) 269-426-3744; Robert.Tatina@dwu.edu
Webmaster	Tim Mullican, DWU, biology 995-2715, Tim.Mullican@dwu.edu
Member-at-Large 2023-2025	Mark Gabel, BHSU biology (retired) 642-6251, Mark.Gabel@bhsu.edu
Member-at-Large 2023-2025	Abigail Martens, SDSU, Entomology, Insect Biodiversity Lab 688-4593, abigail.martens@sdstate.edu
Member-at-Large 2023-2025	Geore Nora, NSU, Chemistry 626-2471, george.nora@northern.edu
Member-at-Large 2023-2025	Eric Pulis, NSU, Biology 626-2600, eric.pulis@northern.edu
Member-at-Large 2023-2025	Adam Sturlaugson, USF, Chemistry 331-6757, Adam.Sturlaugson@usiouxfalls.edu
Member-at-Large 2020-2024	Arvid Boe, SDSU, Plant Science 688-4759, Arvid.Boe@sdstate.edu
Member-at-Large 2020-2024	Bruce Cuevas, USD, Basic Medical Sciences 658-6330, Bruce.Cuevas@usd.edu
Member-at-Large 2020-2024	Paul Johnson, SDSU, Entomology 688-4438, paul.johnson@sdstate.edu
Member-at-Large 2020-2024	Xu Lan, SDSU, Department of Natural Resource Management 688-4564. Lan.Xu@sdstate.edu
Member-at-Large 2020-2024	Chao Zhang, MMU, Physics 668-1588, chao.zhang@mountmarty.edu
Graduate Student Member-at-Large, 2023-2025	open

FELLOWS OF THE SOUTH DAKOTA ACADEMY OF SCIENCE 1998-2022

1988	Kenneth F. Higgins (SDSU)
1999	The late Chuck Estee (USD)
2001	Carroll Hanten
2001	Emil F. Knapp (AU)
2003	The late Clyde Brashier (DSU)
2003	The late Milton Hanson (AU)
2003	The late Sr. Laeticia Kiltzer (MMC)
2003	The late Robert Looyenga (SDSM&T)
2003	Arlen Viste (AU)
2003	The late Everett White (SDSU)
2004	Perry H. Rahn (SDSM&T)
2004	Robert Stoner (USD)
2011	Maureen Diggins-Hutcheson (AU)
2011	Gary Earl (AU)
2011	The late H.L. Hutcheson (SDSU)
2011	Wally Klawiter Jr. (USF)
2012	Nels Granholm (SDSU)
2012	The late Audrey Gabel)BHSU)
2012	Mark Gabel (BSU)
2012	Robert Tatina (DWU)
2015	Arvid Boe (SDSU)
2015	R. Roy Kintner (AU)
2015	The late Gary Larson (SDSU)
2015	The late Norm Miller (SDSM&T)
2015	David Parris (NJ Museum)
2016	Donna Hazelwood (DSU)
2016	Cliff Lewis (USF))
2016	James Martin (SDSM&T)
2016	R. Neil Reese (SDSU)
2017	Dave Bergmann (BHSU)
2017	Andrew Detwiler (SDSM&T)
2019	Nels Troelstrup (SDSU)
2020	Larry Stetler (SDSM&T)
2020	Duane Weisshaar (AU)
2021	Mile Koppang (USD)
2021	Paul Johnson (SDSU)
2022	Jeffrey Palmer (DSU)
2022	David Swanson (USD)

HONOR ROLL OF SOUTH DAKOTA ACADEMY OF SCIENCE PAST PRESIDENTS

prelim	1914-1915	†Hilton Ira Jones, Chemistry, Dakota Wesleyan University
1st	1915-1916	†Hinton Ira Jones, Chemistry, Dakota Wesleyan University
2nd	1916-1917	†A. N. Cook, Chemistry, University of South Dakota
3rd	1917-1918	†A. N. Hume, Agronomy, State College of Agriculture and Mechanic Arts
4th	1918-1919	†J. Gladden. Hutton, Soils, State College of Agriculture and Mechanic Arts
5th	1919-1920	†Freeman Ward, Geology, University of South Dakota
6th	1920-1921	†Charles W. Cuno, Chemistry, Yankton College
7th	1921-1922	†Doane Robinson, Historian, Pierre
8th	1922-1923	†A .N. Hume, Agronomy, State College of Agriculture and Mechanic Arts
9th	1923-1924	†Arthur M. Pardee, Chemistry, University of South Dakota
10th	1924-1925	†L. A. Stout, Mathematics, Dakota Wesleyan University
11th	1925-1926	†J. Gladden, Hutton, Soils, State College of Agriculture and Mechanic Arts
12th	1926-1927	†A. P. Larrabee, Biology, Yankton College
13th	1927-1928	†Homer Hill, Science, Huron College
14th	1928-1929	†H. C. Severin, Entomology, State College of Agriculture and Mechanic Arts
15th	1929-1930	†E. P. Rothrock, Geology, University of South Dakota
16th	1930-1931	†V. A. Lowry, Science, Eastern State Teacher's College
17th	1931-1932	†B. B. Brackett, Communications, University of South Dakota
18th	1932-1933	†George Gilbertson, Entomology, State College of Agricult and Mechanic Arts
19th	1933-1934	†A. V. Arlton, Biology, Dakota Wesleyan University
20th	1934-1935	†E. P. Churchill, Biology, University of South Dakota
21st	1935-1936	†William H. Powers, Librarian, State College of Agriculture and Mechanic Arts
22nd	1936-1937	†Greg M. Evans, Chemistry, Yankton College
23rd	1937-1938	†Charles A. Hunter, Bacteriology, University of South Dakota
24th	1938-1939	†Ward L. Miller, Botany, State College of Agriculture and Mechanic Arts
25th	1939-1940	†Walter V. Scaright, Geology, University of South Dakota
26th	1940-1941	†A. L. Moxon, Chemistry, State College of Agriculture and Mechanic Arts
27th	1941-1942	†Henry Lowsma, Science, General Beadle State College
28th	1942-1943	†Edwin H. Shaw, Jr., Biochemistry, University of South Dakota
29th	1943-1944	†Orin M. Lofthus, Biology, Augustana College
30th	1944-1945	†Lester S. Guss, Chemistry, South Dakota State College
31st	1945-1946	†Raymond J. Greb, Biology, Huron College
32nd	1946-1947	†F. L. Bennett, Botany, Black Hills Teachers College
33rd	1947-1948	†A. L. Haines, Chemistry, University of South Dakota
34th	1948-1949	†V. S. Webster, Chemistry, South Dakota State College
35th	1949-1950	†John M. Froemke, Chemistry, Augustana College
36th	1950-1951	†William E. Ekeman, Mathematics, University of South Dakota
37th	1951-1952	†E.R. Binnewies, Chemistry, South Dakota State University
38th	1952-1953	John M. Winter, Botany, University of South Dakota
39th	1953-1954	H. R. Fossler, Psychology, University of South Dakota
40th	1954-1955	†Harlan L. Klug, Chemistry, South Dakota State University
41st	1955-1956	†Frank W. Jobes, Biology, Yankton College
42nd	1956-1957	†Charles R. Estee, Chemistry, University of South Dakota
43rd	1957-1958	†John Willard, Chemistry, South Dakota School of Mines & Technology
44th	1958-1959	Shapley W. Howell, Science, Yankton College
45th	1959-1960	†V. R. Nelson, Physics and Aeronautics, Augustana College
46th	1960-1961	†Ernest J. Hugghins, Parasitology, South Dakota State University

47th	1961-1962	†Charles M. Vaughn, Science, University of South Dakota
48th	1962-1963	†Morton Green, Biology, South Dakota School of Mines & Technology
49th	1963-1964	†S. G. Froiland, Biology, Augustana College
50th	1964-1965	†George P. Scott, Chemistry, University of South Dakota
51st	1965-1966	†Marvin Hanson, Mathematics, Huron College
52nd	1966-1967	†Robert Sandvig, Chemistry, South Dakota School of Mines & Technology
53rd	1967-1968	†Walter Morgan, Biology, South Dakota State University
54th	1968-1969	†Charles Sidlo, Physics, Sioux Falls College
55th	1969-1970	†C. L.Hills, Biology, Dakota Wesleyan University
56th	1970-1971	R. Roy Kintner*, Chemistry, Augustana College
57th	1971-1972	†Raymond D. Dillon, Zoology, University of South Dakota
58th	1972-1973	†Veronica Fasbender, Biology, Mount Marty College
59th	1973-1974	David Ballew, Mathematics, South Dakota School of Mines & Technology
60th	1974-1975	†Jack K. Saunders, Biology, Northern State University
61st	1975-1976	†Harold Foss, Natural Science, Augustana College
62nd	1976-1977	Waldemar F. Klawiter, Jr.*, Mathematics and Physics, Sioux Falls College
63rd	1977-1978	†James C. Schmulbach, Biology, University of South Dakota
64th	1978-1979	James A. Fries, Chemistry, Northern State University
65th	1979-1980	†John Landegent, Biology, Sioux Falls High School
66th	1980-1981	†Norman E. Miller, Chemistry, University of South Dakota
67th	1981-1982	Harry G. Hecht, Chemistry, South Dakota State University
68th	1982-1983	Theodore Van Bruggen, Botany, University of South Dakota
69th	1983-1984	L. Cliff Lewis*, Chemistry, Sioux Falls College
70th	1984-1985	Maureen Diggins-Hutcheson*, Reproductive Biology, Augustana College
71st	1985-1986	Don Hopkins, Physics, South Dakota School of Mines & Technology
72nd	1986-1987	Stephen R. Metzner, Science, University of South Dakota
73rd	1987-1988	†Marvin Selnes, Science, Sioux Falls School District
74th	1988-1989	Warren Hein, Physics, NSU/SDSU
75th	1989-1990	James Martin*, Paleontology, South Dakota School of Mines & Technology
76th	1990-1991	Samuel Gingerich, Chemistry, Northern State University
77th	1991-1992	Mark Gabel*, Botany, Black Hills State University
78th	1992-1993	Nels Granholm*, Biology and Global Studies, South Dakota State University
79th	1993-1994	Keith Perkins, Biology, Sioux Falls College
80th	1994-1995	Arvid Boe*, Agronomy, South Dakota State University
81st	1995-1996	Tim Sorenson, Mathematics, Augustana College
82nd	1996-1997	John Thomas, Biology, University of South Dakota
83rd	1997-1998	Sharon Clay, Plant Science, South Dakota State University
84th	1998-1999	Royce Engstrom, Chemistry, University of South Dakota
85th	1999-2000	Neil Reese*, Botany, South Dakota State University
86th	2000-2001	Lenore Koczon, Chemistry, Northern State University
87th	2001-2002	Charles Lamb, Biology, Black Hills State University
88th	2002-2003	†Steve McDowell, Chemistry, South Dakota School of Mines & Technology
89th	2003-2004	Miles Koppang*, Chemistry, University of South Dakota
90th	2004-2005	Andy Detwiler*, Atmospheric Sciences, S D School of Mines & Technology
91st	2005-2006	Robert Tatina*, Biology, Dakota Wesleyan University
92nd	2006-2007	James Sorenson, Biology, Mount Marty College
93rd	2007-2008	Michael Wanous, Biology, Augustana College
94th	2008-2009	Nels H. Troelstrup, Jr., Biology, South Dakota State University
95th	2009-2010	Krisma DeWitt, Chemistry, Mount Marty College
96th	2010-2011	Dave Bergmann*, Biology, Black Hills State University
97th	2011-2012	Krisma DeWitt, Chemistry, Mount Marty College

98th	2012-2013	†Gary Larson*, Biology, South Dakota State University
99th	2013-2014	†Gary Larson*, Biology, South Dakota State University
100th	2014-2015	Chun Wu, Chemistry, Mount Marty College
101st	2015-2016	Steven Matzner, Biology, Augustana University
102nd	2016-2017	George Mwangi, Chemistry, University of Sioux Falls
103rd	2017-2018	Vicki Geiser, Northeast Community College
104th	2018-2020	Vicki Geiser, Freeman Regional Hospital
105th	2020-2023	Tim Mullican, Dakota Wesleyan University

*Fellow of the SD Academy of Science
†Deceased

ANNUAL MEETING ROTATION SCHEDULE 2024-2060

Year	Location	Sponsoring Institution
2024	Aberdeen, SD	Northern State University
2025	Rapid City, SD	South Dakota School of Mines and Technology
2026	Sioux Falls, SD	University of Sioux Falls
2027	Mitchell, SD	Dakota Wesleyan University
2028	Brookings, SD	South Dakota State University
2029	Yankton, SD	Mount Marty University
2030	open	open
2031	Spearfish, SD	Black Hills State University
2032	Madison, SD	Dakota State University
2033	Vermillion, SD	University of South Dakota
2034	Aberdeen, SD	Northern State University
2035	Rapid City, SD	South Dakota School of Mines and Technology
2036	Sioux Falls, SD	University of Sioux Falls
2037	Mitchell, SD	Dakota Wesleyan University
2038	Brookings, SD	South Dakota State University
2039	Yankton, SD	Mount Marty University
2040	open	open
2041	Spearfish, SD	Black Hills State University
2042	Madison, SD	Dakota State University
2043	Vermillion, SD	University of South Dakota
2044	Aberdeen, SD	Northern State University
2045	Rapid City, SD	South Dakota School of Mines and Technology
2046	Sioux Falls, SD	University of Sioux Falls
2047	Mitchell, SD	Dakota Wesleyan University
2048	Brookings, SD	South Dakota State University
2049	Yankton, SD	Mount Marty University
2050	open	open
2051	Spearfish, SD	Black Hills State University
2052	Madison, SD	Dakota State University
2053	Vermillion, SD	University of South Dakota
2054	Aberdeen, SD	Northern State University
2055	Rapid City, SD	South Dakota School of Mines and Technology
2056	Sioux Falls, SD	University of Sioux Falls
2057	Mitchell, SD	Dakota Wesleyan University
2058	Brookings, SD	South Dakota State University
2059	Yankton, SD	Mount Marty University
2060	open	open

IN MEMORIUM
DR. THEODORE VAN BRUGGEN

OBITUARY FOR THEODORE VAN BRUGGEN

Dr. Theodore Van Bruggen (97) died on Friday July 28th, 2023, in Pinellas Park, FL. Ted was born on January 26, 1926. He graduated from high school in Ireton, Iowa and went into the U.S. Navy during World War II. In a phone conversation in 2020 Ted told me that he piloted a landing craft during the U. S. invasion of a small Japanese island 75 years prior to the date of our call when he was 19 years old.

Ted married his wife Geraldine Poll on September 30, 1948. After 69 years of marriage Geraldine passed away in 2017. They were the parents of two children, Tom, and Janet.

Ted graduated from Buena Vista University (Storm Lake, Iowa) with a B.S. degree. He earned his M.S. degree from the University of South Dakota (Vermillion) and his Ph.D. from the University of Iowa (Iowa City). His first teaching position was at Northwestern College (Orange City, Iowa). He and Geraldine later moved to Vermillion where Ted was employed by the University of South Dakota (USD). Ted was a professor and department chairman for the Botany Department and later served as Associate Dean of Arts and Sciences as well as other positions.

Dr. Van Bruggen was an active member of the South Dakota Academy of Science. He became a member beginning in 1960 and served as Secretary/Treasurer from 1963-1969. He was on the Academy of Science-NSF Steering Committee from 1963-1967 and participated in various other activities including the Auditing, Resolutions and By-Laws Committees. Ted became a life member in 1977 and was President from 1983-1984.

Ted will forever be known for his pioneering work on the flora of South Dakota. His landmark book *The Vascular Plants of South Dakota* was published in 1976 and was the first real flora of the state. He revised it and published the second edition in 1985. The third edition appeared in 1996. During his professional career he collected well over 7000 plant specimens and contributed greatly to the Herbarium at USD. He also is well-known for his popular photographic guides, the *Wildflowers of the Northern Plains and Black Hills* (1971, 1976) and the expanded and updated *Wildflowers, Grasses and other Plants of the Northern*

Plains and Black Hills (1983, 1992). Ted was also a contributor to the monumental work *Flora of the Great Plains* (1986).

Ted is survived by his son Tom Van Bruggen, Park Rapids, MN and daughter Janet Hazen, Pinellas Park, FL as well as six grandchildren and 12 great grandchildren. His final wishes were to donate his body to the University of South Dakota Sanford Medical School.

Selected Publications

Theodore Van Bruggen. 1950. Comparative rates of root growth of certain native and cultivated grasses. Ph.D. Dissertation. University of Iowa.

Theodore Van Bruggen and J.M. Winter. 1959. Notes on the Flora of South Dakota. Proceedings of the South Dakota Academy of Science. 36:479-82.

Theodore Van Bruggen and J.M. Winter. 1960. Notes on the Flora of South Dakota. Proceedings of the South Dakota Academy of Science. 39: 125-129.

Theodore Van Bruggen. 1961. An Ecologic and Taxonomic Study of a Sand Dune and Flood Plain Area Adjacent to the Missouri River. Proceedings of the South Dakota Academy of Science. 40: 185-198.

Theodore Van Bruggen. 1965. The Cyperaceae of South Dakota. Proceedings of the South Dakota Academy of Science. 44:185-198.

Theodore Van Bruggen. 1967. The Pteridophytes of South Dakota. Proceedings of the South Dakota Academy of Science. 46: 126-144.

Theodore Van Bruggen. 1971. Wildflowers of the Northern Plains and Black Hills. Ed. 1. Badlands Natural History Association, Interior, SD.

Theodore Van Bruggen. 1976. The Vascular Plants of South Dakota. ed. 1. Iowa State University Press.

Theodore Van Bruggen. 1976. Wildflowers of the Northern Plains and Black Hills. ed. 2. Badlands Natural History Association, Interior, SD.

Theodore Van Bruggen. 1980. Rare Native Plants of South Dakota. Proceedings of the South Dakota Academy of Science. 59: 175-183.

Theodore Van Bruggen. 1982. The South Dakota Landscape-Past and Future. Proceedings of the South Dakota Academy of Science. 61: 18-23.

Theodore Van Bruggen. 1983. Wildflowers Grasses and Other Plants of the Northern Plains and Black Hills. ed. 3. Badlands Natural History Association, Interior, SD.

Theodore Van Bruggen. 1985. The Vascular Plants of South Dakota. ed. 2. Iowa State University Press.

The Great Plains Flora Association. 1986. Flora of the Great Plains. University Press of Kansas.

Theodore Van Bruggen. 1992. Wildflowers Grasses and Other Plants of the Northern Plains and Black Hills. ed. 4. Badlands Natural History Association, Interior, SD.

Theodore Van Bruggen. 1996. The Vascular Plants of South Dakota. ed. 3. University of South Dakota Bookstore.

Mark Gabel
Curator of the Herbarium
Black Hills State University
Spearfish, SD

Complete Senior Research Papers

submitted for presentation at the

108th Annual Meeting

of the

South Dakota Academy of Science

NATURAL HISTORY NOTES ON SEED-EATING INSECTS OF AMERICAN LICORICE (*GLYCYRRHIZA LEPIDOTA* PURSH (FABALES: FABACEAE)) IN SOUTH DAKOTA, WITH A NEW SPECIES OF *ACRISIS* FÖRSTER (HYMENOPTERA: BRACONIDAE)

Paul J. Johnson[1], Abigail P. Martens[1], and Arvid Boe[2]
[1]Insect Biodiversity Lab
[2]Department of Agronomy, Horticulture and Plant Science
South Dakota State University, Brookings, South Dakota 57007

ABSTRACT

The known seed-eating insects of American licorice, *Glycyrrhiza lepidota* Pursh, are briefly reviewed with new information and additional observations and interpretations. The associations of the bruchine beetle *Acanthoscelides aureolus* (Horn) and the seed wasp *Bruchophagus grisselli* McDaniel and Boe with licorice are reviewed. The primary parasitoids of *B. grisselli* are the native *Eurydinoteloides incerta* (Ashmead) and the invasive *Mesopolobus bruchophagi* (Gahan). The native host association of the moth *Grapholita interstinctana* (Clemens) with licorice is confirmed, with the larval feeding behavior as a seed predator described and illustrated. A probable larval/pupal parasitoid of this moth is the braconid wasp *Acrisis erwini* **new species**. The true bugs *Alydus eurinus* (Say), *Al. pilosulus* (Herrich-Schaeffer), and *Euschistus variolarius* (Palisot de Beauvois) are newly reported as seed predators of immature licorice seed; **new host records** for each species. *Eurydinoteloides incerta* (Ashmead) is reported associated with *Ac. aureolus* (Horn) as a **new host record**. The moth *Hystrichophora ochreicostana* (Walsingham) is newly associated with licorice as a **new host record**. Known other insects, mostly parasitoids of seed-feeders, associated with licorice are summarized.

Key Words

Seed predator, parasitoid, *Trifolium*, *Grapholita*, *Acanthoscelides*, *Bruchophagus*, *Mesopolobus*, *Eurydinoteloides*, epizoochory, habitat fragmentation, *Alydus*, *Euschistus*, *Hystrichophora*

INTRODUCTION

American licorice, or wild licorice, (*Glycyrrhiza lepidota* Pursh; Fabales: Fabaceae) (Figures 1–3), herein simply as "licorice," is a widespread native forb, often used as an herb, found across temperate northern North America, mostly from the western Mississippi River valley and westward. Though this plant was well known to indigenous peoples for probably thousands of years and to Euroamerican explorers and colonists for a number of decades, it became formally known to western science by Frederick Pursh (1814) who ostensibly described the species from specimens collected on the Lewis and Clark Expedition. In the expedition journal (Lewis *et al.* 2002) for Wednesday, 8 May 1805, Meriwether Lewis wrote, "The wild Licquorice is found on the sides of these hills, in great abundance," the hills being the Milk River Hills on the south side of the Missouri River opposite of the mouth of the Milk River, now in Montana. It was also noted by William Clark that their Shoshone guide Sacagawea gathered roots of licorice during the expedition for food and medicine, and that it was widely used by native peoples of the regions traveled. The species is noted again in the Journal on 20 July 1806 on the return trip as "the wild liquorice and sunflower are very abundant in the plains and river bottoms." Pursh (1814) gave no helpful details beyond the type locality being "on the banks of the Missouri." Further Pursh's notation mentions July and August, which implies specimens in flower and a potential specimen collected during the expedition's return trip, but the description lacks the "*v. s. in Herb. Lewis*" reference to a dried expedition specimen as he normally indicated. Possibly, as Phillips (2003) stated, that no specimen from the Lewis and Clark expedition was actually available for the description due to losses of many botanical specimens. Evidently, other specimens to examine existed from other explorers in the lower Missouri River region. A decade after the expedition, Nuttall (1818) also botanically explored the Missouri River Valley and wrote about licorice as "Abundant around St. Louis, where it was first detected by Mr. John Bradbury, F.L.S.; it is also common on the alluvial banks of the Missouri to the Mountains." As recently as Visher (1914), it was reported that licorice was a "common shrub along stream banks" in some of the least settled parts of the country; see Hoffman and Rogers (1966) and Gabel *et al.* (2010, 2015) for additional vegetational and botanical bibliographies and histories of the region.

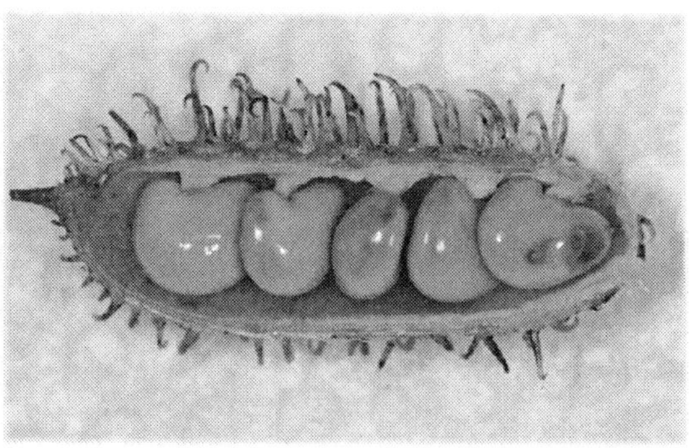

Figure 1. (left) Inflorescences on a single shoot of licorice. Figure 2. (center) Immature mid-season seed pods of licorice on a single shoot. Figure 3. (right) Immature seeds of licorice in natural developmental position within a pod.

Surprisingly for a conspicuous plant with well documented culinary and pharmacological uses (*e.g.*, Rogers 1980; Duke 1981; Morgan and Weeden 1990; Kindscher 1987, 1992; Albers *et al.* 2003; Ode 2006; Wahab *et al.* 2021; Huan *et al.* 2021), there are no studies on its broader natural history. The closely related Eurasian *Gly. glabra* L. shares these traits along with pesticidal and repellent effects from extracts, or biocontrol potential (Xue *et al.* 2022), (*e.g.*, Duke 1981; Wahab *et al.* 2021; Huan *et al.* 2021; Ding *et al.* 2022). Despite all of this there is little recorded entomological investigation on American licorice.

In North America, licorice is reported native from western Ontario to Missouri, and westward (*e.g.*, Britton and Brown 1913; Fernald 1950; GPFA 1986; Kartesz 2015). It is the only native species of the genus on the continent and one of only two in the Americas, the second being *G. astragalina* Gillies ex Hook. & Arn. from Argentina, Brazil, and Chile (Tropicos 2023). Our licorice is occasionally cultivated as an ornamental, but it is not generally used commercially. Though *Gly. glabra* is reported from California, Nevada, and Utah (Kartesz 2015), this is as a managed or escaped cultivar (CalFlora 2023). The USDA Plants Database (2023) illustrates the natural transcontinental distribution of licorice across southern Canada from western Ontario, northern Mexico (Baja California, Chihuahua) (RHNM 2023), and most of the contiguous United States of America, except the southeastern-most states. Kartesz (2015) maps at the county level and shows the recorded presence of licorice in most counties from Minnesota, Iowa, Kansas, Oklahoma and Arkansas, and through most of the western states, with sporadic occurrence eastward from Illinois and Wisconsin to Maine and Virginia, states where it is regarded as introduced since before Fernald (1950). Esser (1994) summarized management information on *Gly. lepidota* in the U.S.A. and listed it from nine ecosystems, 14 physiographic regions, 22 Kuchler plant associations, eight forest cover types, and then eight native and invasive plant communities in the grassland and riparian areas of the northern Great Plains. Kirsch and Kruse (1973) regarded licorice as an "increaser" following grassland fires.

Barghi and Siljak-Yakovlev (1990) reported karyological information at that time on *Glycyrrhiza* species and reported that *Gly. lepidota* shared with *Gly. echinata* (L.) and *Gly. glabra* a chromosome count of 2n = 16 and demonstrated species-level variation in chromosomal morphology. Çetin *et al.* (2014) summarized most of the Eurasian species of *Glycyrrhiza* and found the same chromosome counts, except they also found both 2n = 16 and 2n = 18 for *Gly. pallidiflora*, and 2n = 14 for *Gly. asymmetrica*, a geographically highly restricted species; *Gly. lepidota* was not included in this study. Hayashi *et al.* (2005) separated *Gly. lepidota* and treated it as a sister-group, including *Gly. echinata*, *Gly. macedonica* Boiss. & Orph., *Gly. pallidiflora* Maxim., *Gly. uralensis* Fisch. ex DC., *Gly. inflata* Batalin, and *Gly. glabra*, based on *rbcL* sequences, *i.e.* the rubulose-1,5-bisphosphate carboxylase gene used for species-level relationship analysis. Subsequently, Duan *et al.* (2020) used chloroplast DNA for investigating purported taxon relationships and developed a paleogeographic hypothesis for the distributions of extant species. Their phylogenetic analysis and molecular dating suggest that *Gly. lepidota* is an isolated species of the genus that derived as an early Pliocene migrant via a Beringian land bridge between 5.36–3.5 Ma during a major period of floristic exchange between eastern Asia and North America (*e.g.*, Wen *et al.* 2016). Ancestral relationships seem closest to an eastern Eurasian clade including *Gly. acanthocarpa* (Lindl.) J. Black, *Gly. echinata*, *Gly. pallidiflora*, and *Gly. yunnanensis* P.C. Li. The South American *Gly. astragalina* was treated as a sister-species to *Gly. lepidota* and as a product of long-distance dispersal after about 2.5 Ma by Duan *et al.* (2020).

In South Dakota licorice is found across the state (Johnson and Larson 1999; Larson and Johnson 1999; Boe and Johnson 2016), but is not a species of the drier grass-dominated prairie communities (Stevens 1963). Rather, it is found in mesic well-drained, gravelly, sandy, to light loam soils that are sub-irrigated on lower slopes and side gullies of ravines, swales, and shallow slopes of meadows near sloughs, depressions, ditches, potholes, and erosional benches and terraces above frequent flood zones along streams, ponds, and lakes. Licorice is also found in mesic saline sites in the northern Great Plains (Dodd and Coupland 1966). Licorice nodulates and is a nitrogen fixer (Allen and Allen 1980), grows in moist neutral to slightly alkaline (pH 7.3–7.5) soil conditions, and tolerates light shade to full sun, but does not tolerate hydric or heavy soils, or day-long shade. Andersen and Nelson (2014) regarded it as a "facultatively riparian plant" on nitrogen limited substrates in Arizona and New Mexico. Contemporary occurrences include the former, but also at anthropogenic sites from urbanization and agriculturalization, such as roadsides, roadside ditches, fence lines, mesic fields and pastures, open secondary woodlands, conservation areas, cultural steppe, and a vague partially or undeveloped habitat often termed "waste areas." It could be considered a persistent plant in many disfigured anthropogenic habitats. Within the context of agriculture, licorice grows in so-called "marginal" sites that have below optimum soil and water conditions for row crops.

In the northern Great Plains, licorice typically begins flowering during the last week of June or first week of July. In eastern South Dakota peak flowering is during the third week of July, with most patches reaching 90% bloom at this time regardless of actual latitude or longitude. However, flowering can range into mid-September on late developing shoots, or lateral shoots induced by loss of apical portions of the main growth. The flowers are hermaphroditic with 20–30 flowers per raceme (Matteson 1933; PJJ and AB pers. observ.) and are apparently pollinated primarily by bees (Hymenoptera: Apoidea). Hobbs (1966) noted bumble bees (*Bombus* spp.) as pollinators in Alberta. Hurd (1979) listed 18 species of bee taken at licorice, mostly in California; pollinators in the Great Plains require study.

The inflorescence is upright and blooms from the base to apex, often taking 4–5 weeks for all flowers on a shoot to complete development. Fertilized flowers will form pods (botanically these are fruits, called legumes but we use the conventional term) normally containing 1–7 seeds (3–5 seeds; Stevens 1963), smaller pods may have only one or two seeds, and larger pods may have 5–7 seeds. There is a tendency for smaller size and fewer seeds in the proximal and distal pods, and larger pods with more than 4–6 seeds in pods of the midlength third to half of the raceme. The pods also develop from base to apex, are green when immature, changing to dark brown at maturity and are armed with recurved or hooked prickles. It is typical that the lower flowers senesce and pods form before more apical buds open. All of the pods are indehiscent as noted by Johnson (1981) and Boe and Wynia (1985), thus providing protection for seed-eaters and associates until pod decay in the spring.

Flowering by licorice is usually completed by the end of the first week of August, except terminal clipping events, such as deer browsing, may induce lateral bud growth and produce late season flowers. The seed within the pod is positioned with the funiculus attached internally to the placenta that parallels the adaxial suture. The pod is positioned more or less perpendicularly from the rachis when immature, with the apex directed downward at maturity and the suture adaxial, or dorsad. Mature pods are dry and the prickles stiff, with both integument and prickles softening and becoming flexible with moisture from rain or snow. The pods are persistent with 70–80% falling during winter, though the remainder may remain on the expired shoot until spring.

Although plants investigated by us were generally in the 6–8 dm height range, individual shoots did reach about 1.0 m, with plants characteristically spreading extensively by rhizomes, often forming large patches of intertwined root systems. Boe and Wynia (1985) found that one-year old plants in tilled plots originally planted 100 cm apart had rapid root growth and entanglement preventing recognition of individual plants. The root system can be complex and deep, with Weaver (1954: Figure 40, 1968: Figure 9) noting roots extending 4–5 m deep, which makes licorice among the five most deeply rooted native prairie plants. The species was evaluated as good for mine-spoils revegetation (Bjugstad and Whitman 1984, 1989).

Weaver (1954) noted that "Wild licorice is greedily eaten by livestock, especially when it is young." The plant becomes less appealing to livestock as it matures and is apparently not commonly taken under light to moderate grazing regimes when better forage is available. It will be "excluded in heavily grazed areas" (Weaver 1968; Johnson and Nichols 1970). The pods are epizoochoric, *i.e.* distribute by adhering to animals, and though are not problematic for large wildlife, they can cause serious economic quality problems when entangled in sheep wool (SDDA 2009). The plant does not tolerate trampling, intensive grazing regimes, invasive weeds, mowing, haying, herbicides, treeing, or other continuous or frequent impacts on above-ground growth. Given the regional history of landscape changes, the plants now tend to occur in patches widely separated due to ecological conversions in South Dakota, and that they seem not readily colonized by associated insects suggests a pattern of "extinction of ecological interactions" (Janzen 1974) in a fragmented habitat landscape. The plants' persistence is likely due to it being well adapted to areas not suitable for conventional crop production due to poor drainage and salinity. Where it persists, it provides biodiversity enhancement, increased forage variety, production and quality, improved soil fertility and carbon capture and storage, wildlife food and habitat, and general ecosystem goods and services.

As noted by Boe and Johnson (2016), it is a valuable conservation species and used by a variety of wildlife from predatory and small mammals to granivorous birds and insects. In our region it is browsed by large native mammals such as bison (*Bison bison bison* L.), elk (*Cervus canadensis* Erxleben), mule deer (*Odocoileus hemionus* Rafinesque), white-tail deer (*O. virginianus* Zimmerman), and pronghorn (*Antilocapra americana* Ord). The roots are eaten by pocket gophers (*Thomomys* spp.), and the epizoochoric seed pods are naturally spread by at least bison, elk, deer, and racoon (*Procyon lotor* L.) (*e.g.*, Berthoud 1892; Kulbaba *et al.* 2009; Eyheralde 2015). Licorice is a frequent species around bison wallows. Discussions of the role of Pleistocene megafauna species on grassland evolution (*e.g.*, Willerslev *et al.* 2014; Bråthen *et al.* 2021) allows speculation on whether licorice epizoochory was associated with these extinct ungulates given the minimum 3.5 Ma age estimate for the species in North America (Duan *et al.* 2022) and its relationship to *Gly. echinata* and migration from central Asia.

The known insects associated with licorice are largely restricted to reports of seed predation and seed predator parasitoids in the Great Plains (Baker 1895; Boe *et al.* 1981; Boe and Wynia 1985, Boe *et al.* 1988; McDaniel and Boe 1991b; Boe and Johnson 2015, 2016; Martens 2019). This granivory guild is best known by *Acanthoscelides aureolus* (Horn) (Coleoptera: Chrysomelidae: Bruchinae) in regions east of the Rocky Mountains and *Ac. fraterculus* (Horn) in regions of the mountain states and westward (Johnson 1970). The granivorous wasp *Bruchophagus grisselli* McDaniel and Boe (1991b) (Hymenoptera: Eurytomidae) is documented from licorice seed in North Dakota and South Dakota, and Boe and Johnson (2016) suggested licorice as the native host of the seed-eating caterpillar of the leaf-rolling moth *Grapholita interstinctana* (Clemens) (Lepidoptera: Tortricidae).

Our objectives here are to add new observations on the known licorice seed-eating insects; confirm licorice as a native host for *Gr. interstinctana* otherwise known as a forage crop pest; describe a new species of *Acrisis* Foerster (Hymenoptera: Braconidae) associated as a larval/pupal parasitoid of *Gr. interstinctana*; report *Eurydinoteloides incerta* (Ashmead) (Hymenoptera: Pteromalidae) as an *Ac. aureolus* parasitoid for the first time; report the broad-headed bugs *Alydus eurinus* (Say) and *Al. pilosulus* (Herrich-Schaeffer) (Hemiptera: Alydidae) and the stink bug *Euschistus variolarius* Palisot de Beauvois (Hemiptera: Pentatomidae) as feeders on immature licorice seed; and report *Hystrichophora ochreicostana* (Walsingham) (Lepidoptera: Tortricidae) as associated with licorice as a possible granivore or herbivore.

METHODS

This study is an extension of studies by Boe and Johnson (2016), who reported on licorice seed-eaters from sites mostly in western South Dakota, and Martens (2019). In 2022 and 2023 especially, populations of licorice were sought throughout eastern South Dakota with a few new supplementary sites in western South Dakota. Over the past 35 years or so, we now have accumulated observations and data from 65 sites (Figure 4) where each licorice population appeared to be at least 3 years of age, but generally older. Pod samples were usually taken following two weeks of initial development and continued through their maturation and into autumn, with occasional samples during winter. Samples were randomly collected within and around colonies of plants. Each sample consisted of entire inflorescences with approximately 15–25 pods each and at varying stages of maturity, with a sampling range of approximately 100–800 pods per location depending on plant patch size and pod abundance. Pods were placed into and retained in zip-style plastic bags, hung from cords indoors where they experienced morning sunlight only and ambient room temperatures. Each sample bag was monitored, usually daily, for emergence and activity of seed-eaters and parasitoids. Excessive moisture inside bags of samples of pre-mature pods was removed as needed by wiping or daubing with paper towels and the bags left open for several hours each day until dry. Mature pods were air dried for several days before bag closure. Pods were separated and mixed for subsequent random selection. Subsamples of pods were randomly selected, examined and dissected in search of seed-eating and other insects to discover or monitor development of larvae of *Ac. aureolus*, *B. grisselli*, or parasitoids and other associates. Immature and near mature pods were examined shortly after their field collection for *Gr. interstinctana* larvae, or cocoons and pupae. Observations on *Alydus* spp. and *Eus. variolarius* done in the field and lab were mostly on probing and feeding behavior.

Pod and seed sampling of licorice was evaluated based on number of seeds, number of seeds eaten by target taxa, number of seeds fed upon by hemipteroid seed-feeders, and developmentally aborted. The number of larvae per pod and per inflorescence were recorded. Mature pods were collected in late summer through autumn and stored in zip-style plastic bags with insects removed until reduction of emergence frequency, then pods and seeds dissected for overwintering stages. Each pod was examined externally for entry or emergence holes, and the suture split to reveal the seeds and insect feeding damage. Each seed was counted and

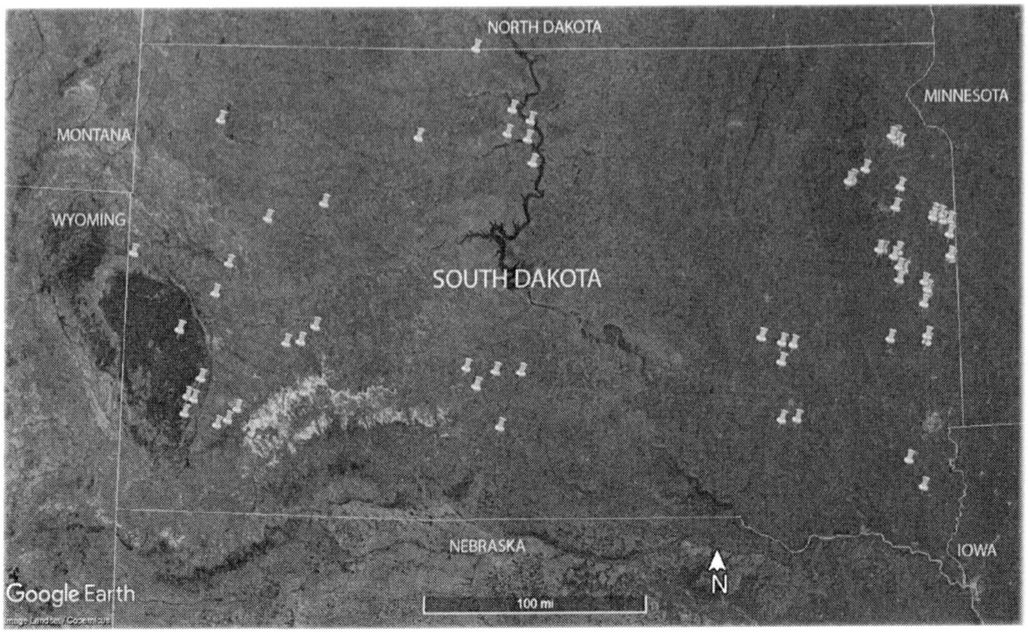

Figure 4. Map of South Dakota with pins showing locations of licorice samples.

evaluated for viability, insect damage, and potential insect occupancy, and then dissected for status, species identification, and damage characteristics.

Viable seeds have a shiny, green-toned testa and feel solid to the touch with fingers or forceps. Seeds with *Ac. aureolus* larvae are soft to the touch, and the testa either dull green or dark brown colored. Seeds from which *Eury. incerta* or *Mesopolobus bruchophagi* (Gahan) (Hymenoptera: Pteromalidae) (Figure 4) emerged had a ragged-edged emergence hole, usually on a side. Seeds from which *Ac. aureolus* emerged had a clean-edged round hole in the side or basal end of the seed that typically matched a similar sized hole penetrating the valve. Seeds fed upon by *Alydus* spp. or *Eus. variolarius* were mottled in color or shriveled and brown. Feeding by *Gr. interstinctana* larvae damaged or consumed all the seeds in a pod and deposited extensive amounts of feces that often molded. Seed damage not attributed to any of these insects was regarded as "Other", and this category includes seeds aborted for unknown reasons.

Summer samples of mature pods provided early emerging individuals of *Ac. aureolus* and *M. bruchophagi* in high numbers during autumn, and small numbers of other seed-eaters and parasitoids emerging in low numbers through winter and spring. Dry mature pods and seeds were softened in a humidity chamber for a few days or set into a bowl of hot tap water and cooled to room temperature for 1-2 hours to soften tissues to ease dissection and minimize damage to immature insects. Field sampling from late autumn to early spring provided pods with softened integuments that were easy to dissect.

Adults of *Gr. interstinctana* active on licorice and other plants during daylight periods of the summer were observed resting or conducting their circular and reversal defensive dance on leaves of various plants. Example larvae collected from within pods were dispatched at near-boiling water temperatures and preserved in 90–95% ethyl alcohol in 2-dram glass vials. Most undamaged larvae were allowed to complete development, with occupied cocoons individually kept in small petri dishes set within a lidded plastic box and overwintered at ambient temperatures in an unheated garage space. Pupal exuviae and cocoons were collected and kept dry in 2-dram glass vials. Adults were collected and frozen, then mounted and air-dried.

Pods subsampled for occupant evaluation were taken from the 11 sites given in Table 1. Location data for each site reported, all in South Dakota, are: 175th Street south of Watertown, Codington Co., 44.8477N, -97.0367W; Lake Co., 44.1023N, -97.3407W; Rasmussen 2-2, Codington Co., 45.0357N, -97.4311W; Innovation Campus, Brookings Co., 44.3204N, -96.7672W; 158th Street @ 445th Avenue, Codington Co., 45.0941N, -97.3044W; 4 mi. E. Clear Lake, Deuel Co., 44.7532N, -96. 6021W; Six Mile Creek, Brookings Co., 44.4440N, -96.6502W; Felt Farm, Brookings Co., 44.3693N, -96.7961W; Summit Lake GPA, Grant Co., 45.2700N, -97.0101N; McNenny Hatchery, Lawrence Co., 44.5592N, -104.0148W; Lake Poinsett State Park, Brookings Co., 44.5414N, -97.0692W.

The *Acrisis erwini* new species description is the work of author APM. Previously published information and data on the taxonomy and biology of seed-eating and parasitoid wasps were taken primarily from references cited in Muesebeck et al. (1951), Peck (1963), Krombein *et al.* (1979), and Noyes (2019).

Above and below, to distinguish similar genus abbreviations, we used frequently the first two or three letters of the name a instead of only the initial one: *Acanthoscelides* is "*Ac.*", *Alydus* is "*Al.*", *Astragalus* is "*As.*", *Eupelmus* is "*Eup.*", *Eurytoma* is "*Eur.*", *Eurydinoteloides* is "*Eury.*", *Euschistus* is "*Eus.*", *Gleditsia* is "*Gle.*", *Glycyrrhiza* is "*Gly.*" and *Grapholita* is "*Gr.*"

Table 1. Viable, seeds appear shiny, firm, smooth; *A. aureolus* larva or adult in seed; *B. grisselli* as live or identifiable larvae; *M. bruchophagi* adults, larvae or seeds with attributable damage; Eur. tylodermatis larvae or adults in seeds; Alydus/Euschistus, seeds damaged by feeding or probing by true bugs; Grapholita, seeds consumed by caterpillar; Other, abortion or mortality by inconclusive mode.

Site	Pods	Seeds	Viable	*Acanthoscelides aureolus*	*Mesopolobus bruchophagi*	*Bruchophagus grisselli*	*Eurytoma tylodermatis*	*Alydus/ Euschistus*	*Grapholita interstinctana*	Other
175th S. of Watertown	49	122	25	26	0	0	0	17	44	8
Lake Co.	37	154	140	0	0	0	0	5	0	9
Rasmussen 2-2	53	177	3	77	15	0	0	13	63	13
Innovation Campus	67	204	16	15	0	0	13	14	32	15
158th @ 445th	55	164	137	0	0	0	0	43	0	4
4 mi E. Clear Lake	56	210	40	7	0	0	0	132	28	1
Six Mile Creek	60	170	32	60	0	0	0	40	38	0
Felt Farm	17	31	15	0	0	0	0	17	9	0
Summit Lake GPA	79	227	49	50	3	0	0	44	76	3
McNenny Hatchery	91	178	32	29	20	10	3	38	73	22
Lake Poinsett	42	125	9	26	0	0	0	31	62	0
Total	**586**	**1752**	**498**	**308**	**57**	**10**	**19**	**423**	**425**	**75**
Percentage			**27.6%**	**17.1%**	**3.2%**	**0.6%**	**1.1%**	**23.4%**	**23.5%**	**4.2%**
Average	**53.3**	**160.2**	**45.3**	**28.0**	**5.2**	**0.9**	**1.7**	**38.5**	**38.6**	**6.8**

Table 2. Insects reared or collected from licorice in South Dakota. Compiled from Boe and Wynia (1985), Boe et al. (1988), Boe and McDaniel (1991), McDaniel and Boe (1991b), Boe and Johnson (2016), and this report.

Taxon	Ecological Role
Coleoptera	
Chrysomelidae: Bruchinae	
Acanthoscelides aureolus (Horn)	seed predator of *Gly. lepidota* in eastern North America
Acanthoscelides fraterculus (Horn)	seed predator of *Gly. lepidota* in western regions
Cleridae larva; poss. *Enoclerus rosemarus* (Say)	predator of *Gly. lepidota* visitors; larva in empty pods
Hemiptera	
Alydidae	
Alydus eurinus (Say)	feeds on green seeds
Alydus pilosulus (Herrich-Schaeffer)	feeds on green seeds
Aphididae	
Aphis craccivora Koch	feeds on terminal growth, flowers, green pods
Pentatomidae	
Euschistus variolarius (Palisot de Beauvois)	feeds on green seeds
Hymenoptera	
Braconidae	
Acrisis erwini Martens sp. n	presumed parasitoid of *Gr. interstinctana*
Eurytomidae	
Bruchophagus grisselli McDaniel & Boe	seed predator of *Gly. lepidota*
Eurytoma tylodermatis Ashmead	parasitoid of either *Ac. aureolus* or *B. grisselli*
Eupelmidae	
Eupelmus vesicularis (Retzius)	parasitoid of *Ac. aureolus*
Ichneumonidae	
Gelis sp.	presumed parasitoid of *Gr. interstinctana*
Trichogrammatidae	
Uscana semifumipennis Girault	parasitoid of *Ac. aureolus* eggs (Boe and McDaniel 1991)
Pteromalidae	
Dinarmus acutus Thompson	parasitoid of *Ac. aureolus* and *B. grisselli*
Eurydinoteloides incerta (Ashmead)	parasitoid of *Ac. aureolus*
Mesopolobus bruchophagi (Gahan)	parasitoid of *B. grisselli*
Torymidae	
Idiomacromerus perplexus Gahan	parasitoid of *B. grisselli*
Idiomacromerus terebrator Masi	parasitoid of *B. grisselli*
Lepidoptera	
Tortricidae	
Grapholita interstinctana (Clemens)	seed predator
Hystrichophora ochreicostana (Walsingham)	foliage or seed predator
Sparganothis sulfurana (Clemens)	likely only an occasional leave roller

Table 3. Other insects, except pollinators, reported from licorice or licorice predators.

Hemiptera	
Miridae	
Lopidea dakota Knight	a polyphagous plant bug; Asquith (1990), Scudder (1997)
Hymenoptera	
Braconidae	
Agathis annulipes (Cresson)	parasitoid of *Gr. interstinctana*; Allen (1962)
Bassus dimidiator (Nees)	parasitoid of *Gr. interstinctana*; Folsom (1909)
Bracon mellitor Say	parasitoid of *Gr. interstinctana*; Folsom (1909)
Ichneumonidae	
Phanerotoma fasciata Provancher	parasitoid of *Gr. interstinctana*; Allen (1962)
Sphelodon phoxopteridis (Weed);	parasitoid of *Gr. interstinctana*; Marsh and Carlson (1979)

Table 4. Summary of predominate licorice feeders in South Dakota and their parasitoids as number of species by family.

Insect	Feeding mode	Parasitoids
Acanthoscelides aureolus	mature seeds, chewer	1 Trichogrammatidae 1 Eupelmidae 1 Eurytomidae 2 Pteromalidae (1 shared)
Alydus & *Euschistus*	green seeds, fluid extractors	unknown
Bruchophagus grisselli	mature seeds, chewer	2 Torymidae 3 Pteromalidae 1 Eurytomidae 2 Pteromalidae (1 shared)
Grapholita interstinctana	green seeds, chewer	4 Braconidae 3 Ichneumonidae
Hystrichophora ochreicostana	leaves? chewer	undet. Tachinidae

RESULTS AND DISCUSSION

Results of pod, seed examinations and damage causation counts for 2023 samples are given in Table 1. Discussion of each seed-eater species and associates follow. Table 2 provides a summary of insects consistently reared or collected from licorice in South Dakota, Table 3 lists other non-pollinator insects reported from licorice, and Table 4 summarizes the parasitoid numbers and family-level identity for each taxon investigated.

Acanthoscelides aureolus (Horn), licorice seed beetle
(Coleoptera: Chrysomelidae: Bruchinae) (Figures 5–6)

Acanthoscelides aureolus, or licorice seed beetle, may be the most frequently encountered seed-eater of licorice in our samples, due to its conspicuousness. In our 2023 samples (Table 1), *Ac. aureolus* was the third most impactful seed predator on licorice in about 17% of seeds damaged or destroyed and is found at nearly all sampled occurrences of the host throughout the state, including many small, new or relictual populations.

The first report of a bruchine seed beetle on licorice seems to be Baker (1895) who listed *Ac. fraterculus* (Horn), as "*Bruchus fraterculus*", from a plant in Colorado. Johnson (1970) gave *Ac. aureolus* (Horn) as the bruchine associated with *Gly. lepidota* in the midwestern and Great Plains states and provinces, among a long list of *Astragalus* species, and noted records of *Ac. fraterculus* from the same plants in western regions. Owen (1991) added *A. kentrophyta* Gray var. *impexus* (Canby) Barneby to the host list from California. Johnson (1970) discussed that both *Ac. aureolus* and *Ac. fraterculus* are two of 12 species that he regarded as forming his "*Ac. aureolus* complex" and suggested possible cryptic species status for all, being at least partially separated by host plant and geography.

Pfaffenberger and Johnson (1976) studied the larva of *Ac. aureolus* from Arizona populations and found that it morphologically most resembled the larva of

Figure 5. (left) *Acanthoscelides aureolus* male, dorsal aspect. Figure 6. (center) licorice seeds damaged by *Ac. aureolus* larvae, with a larva in situ and a larva removed from seed. Figure 7. (right) *Mesopolobus bruchophagi* female adult.

A. mixtus (Horn). Kingsolver (2004a) included *Gly. lepidota* in a long list of host plants dominated by *Astragalus* spp., with a few other native legumes, for his entry of "*Acanthoscelides aureolus* (Horn) complex." All of the *Acanthoscelides* we reared to date from licorice in South Dakota are referrable to *Ac. aureolus*, based on advice from C. Dan Johnson (*in litt.* to A.B.), the integument and pubescence color, and male genital characters as published by him (Johnson 1970) and fitting the key by Kingsolver (2004a–b); the future use of genomic characters may help to resolve these relationships and associations. Here, we follow Boe and Wynia (1985), Boe *et al.* (1988, 1990), and Boe and Johnson (2016) in the use of *Ac. aureolus* for the taxonomic identity of specimens that we collected or reared from licorice in South Dakota.

We found adult beetles on the plants at flowering, and they were active until pod maturation and early frosts. Boe *et al.* (1990) found that *Ac. aureolus* eggs were laid on pods most frequently near the base and beneath or near to the calyx and that larval entry into green seed was on their edges or upper sides. We also found eggs laid seemingly randomly on the outside of the valves of the immature pods with hatchling larvae burrowing through the chorion, the valve integument, and a tiny hole through the seed testa, usually on or near the chalaza, and into the developing embryo. Entrances through the valve were often identified by the presence of amber colored resin extrusions where the egg or empty chorion is located beneath. In immature pods, it appears that larval burrowing through the valve induces resin deposits on both outside and inside surfaces, much like that made from piercings by true bugs (see *Alydus* and *Euschistus* below). Larval feeding and development are done entirely within a single seed where they consume the embryo and cotyledons, leaving only powdery residue, fecal materials, and ecdysial fragments packed into the distal one-third of the seed beneath the chalaza. Careful dissection of a green occupied seed finds a thin and shallow etching in the tissues from the pin-hole entry and leading to the embryo where a tiny accumulation of feces, exuvial fragments, or a dead larva may be found. Larval mortality may be the cause of what appears as aborted seeds and entered in Table 1 as "Other," and may suggest anti-feeding toxicants in the early embryo (Janzen 1969). With survivorship of *Ac. aureolus* in licorice it seems any toxicants may either disappear at embryo maturity or a large fraction of the larval population may overcome the effects. The post-feeding mature larva was positioned with its head directed proximally and face dorsally within the hollowed testa, and with the abdominal apex toward the funiculus attachment. Winter was spent in quiescent torpor or diapause until mid-spring, usually as a mature larva, though a small fraction of the population pupates during the winter and awaits spring as a pharate adult. Pupation was within the same seed, and the pupa positioned much like the mature larva. The emerging adult chewed through the testa and the mature valve for summer emerging beetles, with overwintering individuals awaiting spring and emerging from the weathered and fragmented pods.

In pod surveys reported by Boe *et al.* (1990), seed predation levels among 25 populations ranged from zero to 92%, averaging about 47% for 17 of those populations. Boe and Johnson (2016) reported nearly 70% of pods had at least one seed occupied by *Ac. aureolus,* and more than 47% of all seeds were eaten to some degree. In 2023 (Table 1) we found from nine sites an average of 52.6% of pods had at least one seed occupied by *Ac. aureolus*, with 17% of all seeds eaten with a wide range of seed occupation from 0.0–43.5%, which generally correlates with the age and size of the licorice patch. Ribeiro-Costa and Alameida (2012) observed that *Ac. aureolus* adult size varied according to host seed size, though we found a wide variation in licorice seed beetle size regardless of seed size. Generally, it was observed that frequency of occupation was found low in small colonies of less than 100 shoots and likely only a few years of age, and higher frequencies in large colonies of thousands of shoots, reflecting several decades or more of patch existence, for example. The Innovation Campus patch occupied roughly 150 m^2 and is approximately five years of age, with the plants closely growing and preventing an enumeration of the actual number of plants due to extensive root growth, interweaving, and multiple shoots per rhizome (Boe and Wynia 1985). Here, it was the third year of establishment before *Ac. aureolus* was detected.

Overwintering *Eur. tylodermatis* recovered from subsamples (Table 1) varied considerably. Among two representative subsamples: 1) a mid-December 2023 field from an estimated 350 pods, 2) a subsample of 67 (ca. 20%) randomly selected pods from the Innovation Campus sample contained 204 seeds (x = 3.3 seeds/pod). These showed a 13% occurrence rate of *Ac. aureolus* larvae of which 52% were in diapause for overwintering and 48% were ectoparasitized by *Eur. tylodermatis* as judged by larval size, color, and morphological notes by McDaniel and Boe (1991a). Emergences between 01 September–15 November 2023 were only 26 *M. bruchophagi* (Figure 7). At the Rasmussen #2-2 site, from an estimated 800 pods a subsample of 53 (ca. 7%) randomly selected pods, 44% of seeds were occupied or damaged by *Ac. aureolus*, 9% had *M. bruchophagi* carcasses or remnants. Emergences between 01 September–15 November 2023 were 337 *M. bruchophagi* and six *Eur. tylodermatis*. These numbers are consistent with those noted by Boe and McDaniel (1989) and McDaniel and Boe (1991a).

The first parasitoids reported for *Ac. aureolus* were *Horismenus missouriensis* (Ashmead) (Hymenoptera: Eulophidae) by Hetz and Johnson (1988) in Colorado, and *Microdontomerus anthonomi* (Crawford) (Hymenoptera: Torymidae) by Kingsolver (2004a) and Grissell (2005), though none on licorice. In South Dakota we have not found *H. missouriensis* in licorice, though it is frequent and a parasitoid of the seed beetle *Amblycerus robiniae* (F.) on wild and cultivated honey locust, *Gleditsia triacanthos* (L.). Parasitoids of larvae found were the native *Eur. tylodermatis* Ashmead (Hymenoptera: Eurytomidae) and *Uscana semifumipennis* Girault (Hymenoptera: Trichogrammatidae), and the invasive *Dinarmus acutus* Thomson (Hymenoptera: Pteromalidae) and *Eup. vesicularis* (Retzius) (Hymenoptera: Eupelmidae). Rearings from pods from late August through December (Table 1) typically produced an occasional adult *Ac. aureolus* and large numbers of adult *M. bruchophagi*; no larvae of this wasp were found with overwintering *Ac. aureolus* larvae. *Eurydinoteloides incerta* was found at only two widely disparate sites, Lake Norden in 2022 and McNenny in 2023. Only one *Idiomacromerus perplexus* (Gahan) was recovered. Notably, in 2023 samples, we dissected *Eur. tylodermatis* larvae and *Eury. incerta* adults and reared only one adult *Eup. vesicularis*.

Winter sampling of pods typically found live *Ac. aureolus* larvae, dead first and second instar larvae often shriveled from *M. bruchophagi* feeding, and parasitoid emergence holes in seeds and valves. Carcasses of *M. bruchophagi* are occasionally found within a seed, or within the pod and outside of the seed. *Mesopolobus bruchophagi* evidently has a rapid development as an ectoparasite on *Ac. aureolus* larvae during seed development, and adult wasps begin emerging shortly after maturation of the pods and seeds and continue to emerge until winter with the peak in late September and October. Emergence was generally through a circular hole chewed through the seed testa and directly through the valve. This emergence period and the lack of finding *M. bruchophagi* larvae or live adults in pods and seeds indicated that it overwinters as an adult away from its host. One 2022 sample site also produced a number of *Eury. incerta* from licorice, which we associate with *Ac. aureolus*, a **new host record** (South Dakota, Hamlin Co., Lake Norden, along Hwy eastside of town, 27 August 2022, ex *Glycyrrhiza lepidota* pods, P.J. Johnson; reared 27. viii. 2022–01. xi. 2022; 44 female, 35 male). Twelve specimens of *Eury. incerta* were removed from overwintering seeds collected from the McNenny fish hatchery area. This native wasp has a long native and adventitious host list including both bruchid beetles and weevils (Curculionidae) (Burks 1979) over a broad geographic range in the Americas, including other native prairie legumes of the region (Boe and Johnson 2008; Boe *et al.* 2019).

***Bruchophagus grisselli* McDaniel and Boe, licorice seed chalcid**
(Hymenoptera: Eurytomidae) (Figures 8–9)

Bruchophagus grisselli, or licorice seed chalcid, was one of the major seed-eaters from licorice seeds at various localities in the region during sampling from 1987–1992, and co-occurs with *Ac. aureolus* within pods. Our prior rearing data (unpublished) indicated occasional adult *B. grisselli* may emerge in autumn or winter from pod samples kept indoors, but emergence from pod samples clearly predominates in mid-spring, late April through early June, with a peak in mid

Figure 8. (left) *Bruchophagus grisselli* female adult. Figure 9. (right) *Bruchophagus grisselli* mature larva within broken testa.

and late May. It should be noted that Kingsolver (2004a) did not list *B. grisselli* as an *Ac. aureolus* associate. During the field seasons of 2022 and 2023 we have few observations of the larva, pupa or adult of *B. grisselli* in the field or reared in samples compared to 1986 and 1987. Our 2022–2023 surveys yielded no adults, and searching, sweeping, and beating of plants at differing times of the day during flowering and early and mid-season pod development did not produce adults in the field. In pod samples we found larvae of *B. grisselli* only in samples from McNenny, not from eastern South Dakota (Table 1).

Parasitoids of *B. grisselli* reported are *Idiomacromerus perplexus*, *I. terebrator* and *M. bruchophagi* (Boe and Johnson 2016). Previously, Cushman (1911) listed three parasitoids from "*Bruchus*" *fraterculus*, now *Ac. fraterculus*, the western sister-species to *Ac. aureolus*: *Bracon xanthostigma* Cresson that is now a synonym of *B. mellitor* Say (Marsh and Carlson 1979), a "*Eurytoma*" which is *Eury. tylodermatis* (Burks 1979), and a "pteromalid" which likely is *M. bruchophagi*. The latter is the now widespread and common pteromalid parasitoid of *B. grisselli* as well as the invasive *B. roddi* (Gussakovsky) (formerly as *B. funebris* Howard) on alfalfa, *B. gibbus* (Boheman) on red clover, or *B. platypterus* (Walker) introduced with their host plants.

Comparing recent results to those of prior surveys causes us to speculate whether this native seed-eater is no longer detectable, and why? Two basic hypotheses may be postulated for testing. The first hypothesis relates to the increased fragmentation and reduction in size of licorice patches across the state, especially eastern portions, which may also involve reduced pollinator visitations. These habitat changes relate directly to cropland and pasturage expansion in the region (*e.g.*, Johnston 2014; Wright and Wimberly 2013; Wimberly *et al.* 2017). The second hypothesis involves the cumulative impact of the invasive Eurasian parasitoid *M. bruchophagi*. Boe and Johnson (2016, p. 101) noted this faunal displacement as a "cryptic ecological disaster." In the most recent five years, our continuing surveys yielded fewer *B. grisselli* each year, few of either *Idiomacromerus* species, yet increasingly large numbers of *M. bruchophagi*. For example, in samples from several sites in eastern South Dakota in 2023, samples involving 300-800 pods each in August and September yielded hundreds of autumn-emerging *M. bruchophagi*. The same samples yielded no *Idiomacromerus* and few other parasitoids. *Mesopolobus bruchophagi* is a species known to parasitize *B. platypterus* in trefoil (*Lotus corniculatus* L.) (Ellis and Nang'ayo 1992), and its occurrence in licorice communities may be an indication of parasitoid host-switching or crossover.

Grapholita interstinctana (Clemens), licorice pod caterpillar
(Lepidoptera: Tortricidae: Grapholitini) (Figures 10–11)

Grapholita interstinctana is the most important seed-eater of licorice that we observed or measured (Table 1) in the region, with 23.5% of all seed loss and 38.6% of all examined seeds lost to predation by its caterpillar. In many instances the entire seed content of examined pods was destroyed. We dub it the licorice pod caterpillar. Among the 11 sites sampled in 2023 (Table 1), 23.5% of all seeds

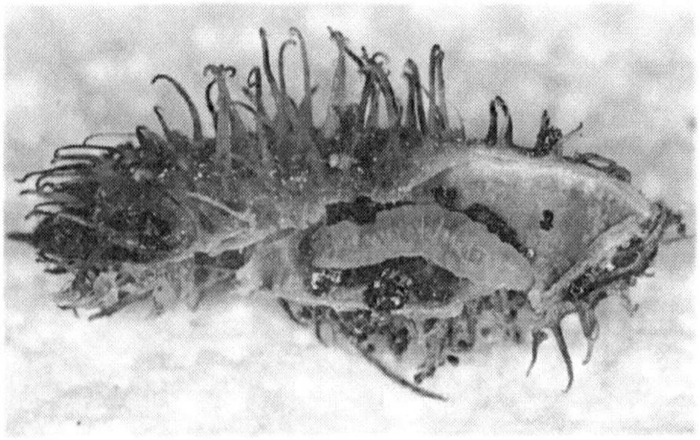

Figure 10. (left) *Grapholita interstinctana* adult resting on leaf. Photo by Thomas Stuart, used with permission. Figure 11. (right) *Grapholita interstinctana* larva inside of predated mature pod.

were predated by this caterpillar, with an average of 38.6 seeds per sample. This is slightly more than the 23.4% of seeds fed on by either an *Alydus* sp. or *Eus. variolarius*, and much more than the 17.1% predation by *Ac. aureolus*.

Clemens (1861) described *Gr. interstinctana* (Clemens), as *Stigmonota interstinctana,* from an unspecified number of adults. No provenance, larva, or host were initially reported. Clemens (1865) distinguished it from *S. tristrigana* Clemens and reported it from Virginia. Grote (1873) redescribed the species as *Grapholitha diastema* and reported it from New York and Pennsylvania, and the misspelling of the genus as "*Grapholitha*" became commonly used thereafter and includes a rather convoluted taxonomic history. Comstock (1881) seems to have provided the first descriptions of feeding damage to red clover (*Trifolium pratense* L.) generalized life history information. Short studies on the insect were commonly done afterwards and tended to repeat the same information, which was summarized by Gossard (1892) and reviewed by Folsom (1909), as *Enarmonia interstinctana*, and Wehrle (1924) as *Laspeyresia interstinctana.* These latter papers summarized the published literature. Interested persons are referred to these for basic life history on this insect on clovers. Heinrich (1926) designated the type as having come from Pennsylvania and gave its distribution as from New Hampshire to Alberta, south to Missouri and North Carolina. Boe and Johnson (2016) reported licorice as a host for *Gr. interstinctana*.

Grapholita interstinctana is a native moth found across northeastern North America south of the boreal forest and east of the Rocky Mountains (Walsingham 1879; Beal 1896; Folsom 1909; Wehrle 1924; Heinrich 1926; Miller 1987; Gilligan *et al.* 2008; Brown 2022) and within the range of licorice east of the Rocky Mountains. It is not known from Eurasia. Historically, the primary or "preferred" host plants are consistently regarded as species and hybrids of *Trifolium* species, especially red clover (*T. pratense*), white clover (*T. repens* L.), alsike clover (*T. hybridum* L.) (*e.g.*, Comstock 1881; Gossard 1892; Lintner 1896; Folsom 1909; Wehrle 1924; Heinrich 1926; Miller 1987; Brown 2022), and crimson clover (*T. incarnatum* L.) (MacKay 1959; Lam *et al.* 2011). This insect remains a pest (Gilligan and Epstein 2014) of forage clovers, though of little concern nowadays except to seed growers, and has been granted several common names with "clover seed caterpillar" being the most used in the past. As a clover pest, this species was familiar to farmers and entomologists at least as early as 1874 (Comstock 1881; Osborn and Gossard 1891a–b). Detailed information on the life cycle and natural history is generally derived from Gossard (1892), Folsom (1909) and Wehrle (1924). Through the early 20th century, clovers became decreasingly important as primary forage for livestock due to increasing farm mechanization, though continued planting and escape moved them across the continent. Yet, red clover is still regarded as a preferred host (Gilligan *et al.* 2008) for the moth.

Brown (2022) listed "*Helianthus* sp." without vetting it as a host for *Gr. interstinctana* as a "seed-feeder in flower head", citing Gilligan *et al.* (2008). However, these latter authors are not referring to seed-eating in *Helianthus*, but are repeating statements from Folsom (1909), Wehrle (1924), and others of activities in clover inflorescences. These latter authors mis-interpreted Gossard (1892) who cited correspondence from F.M. Webster who remarked about the adults being abundant "on the leaves of *Helianthus grosse-serratus*" from Dekalb Co., Illinois in 1877. This activity refers to the sunning and defensive display dancing of the adults on leaves of various plants, which we observed on red clover and cup plant (*Silphium perfoliatum* L.). Likewise, Brown's (2022) listing of *Picea abies* as an implied host based on museum labels is surely referring to sites of collection, likely nearby to populations of one or more clover.

Importantly, none of the historically documented host plants of *Gr. interstinctana* are native to North America. Rather, the clovers noted above were early introductions to North America by colonists with their livestock, other forages, and food crops from Europe. Red clover and other forage clovers were generally grown and used together to feed livestock, particularly draft horses that were the mainstay source of power for tillage, planting, harvesting, transportation, etc., until the development of farm machinery. Afterwards, the decline of livestock for power and the use of clovers as forage and hay mainstays reduced the pest importance of this insect to the point where Manglitz (1985) noted that the insect is now "inconsequential." Red clover has decreased in relative importance as a forage due to the use of insecticides and the limited demand on seed for forage plantings, but it and other clovers are regaining popularity as cover crops and green manures. Red clover continues to increase in distribution and relative abundance as it is now a common plant along roadsides and field edges and has become an important nectar source for pollinators, such as butterflies, moths, and bees, and may be the only abundant nectaring source for late summer and autumn migrating *Danaus plexippus* (L.) and *Vanessa cardui* (L.) (Lepidoptera: Nymphalidae) in agricultural areas of eastern South Dakota. As such, despite about 165 years of pest work on this moth native hosts were not identified until Boe and Johnson (2016) reported the species from seed pods of licorice. Since then, we found larvae within pods and reared adults from licorice at most of the sites indicated in Figure 4.

In our area, *Gr. interstinctana* larvae are found within both green and brown pods from the fourth week of July through August. Larvae appear to begin using licorice pods as soon as basal pods begin expansion. A larva enters a pod at the base, chewing a round hole that is hidden by the floral calyx. Normal positioning of the larval entrance and exit hole and the subsequent frass plug is on the dextral valve, basal and near to the petiole and either central on the valve face or near to the suture. If valve penetration is abandoned or not successful, there is a shallow pit or a short burrow that fails to reach the internal space. On only two occasions were dead first instar larvae found within the burrow.

In a small basal pod, the larva will feed on one to several developing seeds (Figure 11) and then leave the pod to enter a second pod to feed further. A pod that contained a larva will have frass, seed fragments, and remnants of at least one exuvium. The departure of any aged larva is usually through the original entrance hole that is enlarged, or occasionally through an exit hole anywhere along the length of a valve. During feeding the larva will push feces and debris into and through the entrance hole, commonly forming a subcylindrical plug that extends from the pod by 2–3 mm. The presence of the plug will usually indicate the recent or current presence of a larva within the pod. A premature caterpillar may push or chew away the plug and leave the pod for another. Alternatively, either a premature or mature caterpillar may chew a new hole at either the proximal end of the pod, near to the distal end, or rarely at some midlength position. If a premature caterpillar leaves a pod, it will seek entry to a second, or even third, pod to complete its feeding. First feedings, in late July or early August, can occur in larger pods with 4–7 seeds, and these larvae tend to complete development on their first pod. Late season larvae will move to a second pod when there is competition from other seed-eaters or the inside of the pod becomes excessively fouled.

The pupa is stereotypical for a moth, obtect, dark red-brown in color, and enclosed in a thinly woven cocoon. The cocoon is most commonly emplaced between pods, sometimes adhering three pods to form a subtriangular space protected by the rigid setae of the pod or amongst debris within rearing bags. Larvae found in pods collected in late July and early August were judged to be mid-instars, last instars, or prepupae in cocoons. This indicates a single generation per season for our region, which is in contrast to the 2–3 generation per season estimations given by Folsom (1909) and Wehrle (1924) for *Gr. interstinctana* on red clover in more southern and eastern regions.

In the field, only one instance of cocoon spinning and pupation within a pod was observed, and equally rarely were cocoons observed on the outside of pods, and when observed they were amongst the spines and where two or three pods were closely adjacent. This suggests the majority of larvae leave the plant or at least pupate away from pods or pod clusters in which they fed, much like the behavior observed in clover populations (Wehrle 1924). Overwintering is done by the pupa with adults emerging in spring.

Folsom (1909) listed four parasitoids for *Gr. interstinctana* larvae or pupae. In the Braconidae he gave *Bracon mellitor* Say (as *Bracon vernoniae* Ashmead) from Iowa, and *Bassus dimidiator* Nees (as *Microdus laticinctus* Cresson) from Missouri, Iowa, and "Canada" from overwintering cocoons of the moth (Wehrle 1924). From Ichneumonidae, *Phanerotoma fasciata* Provancher (*Phanerotoma* (*Sigalphus*) *tibialis* Haldeman) from the District of Columbia, and *Sphelodon phoxopteridis* (Weed) (as *Glypta leucozonata* Ashmead) from Missouri. *Agathis dimidiator* is a

European introduction; the remainder are apparently native to North America. An undescribed braconid wasp parasitoid reared from *Gr. interstinctana* was recovered from a sample taken in Brookings, SD; it is described here:

***Acrisis erwini* Martens, new species** (Figure 12)

Martens (2019) reported on a collection of a braconid from a licorice sample taken in Brookings, South Dakota, and determined it to be a species of *Acrisis* Förster (Hymenoptera: Braconidae). This species is here formally described and the name validated.

Holotype ♀, U.S.A., South Dakota, Brookings Co. Brookings, 44°18"0.72"N, 96°46'3"W, reared from *Glycyrrhiza lepidota* pods on 12.iii.2018, A. Boe (USNM). Paratypes: 2 ♀, 4 ♂, with the same labels as holotype (USNM, SDSU).

Diagnosis: This species can be separated from *Acrisis americana* by the absence of enlarged hind tibiae in the male; the faint aciculae present on tergite 1 and the anterior third to half of tergite 2 in both sexes; antennae shorter than the combined length of the head, mesosoma, and metasomal T1 and T2; and coloration as discussed below.

Description. Female. Body length 1.6–1.8 mm. *Head.* Eyes large, tentoriocular line equal to malar space, half intertentorial distance, and one-third eye length;

Figure 12. *Acrisis erwini* **Martens, new species, lateral aspect.**

maxillary palpi 5-segmented, labial palpi 3-segmented; antenna 12-segmented, filiform, setiferous; flagellomere 1 (=F1) 4.4 times longer than wide; F2 3.5 times longer than wide and the same width as F1; F5 is 3.1 times longer than wide; F9 and F10 slightly wider than preceding segments; F10 0.7 times the length of F1, tapered toward the tip, lachrymiform; antenna shorter than combined length of head, mesosoma, and metasomal T1 and T2; scape and pedicel combined length 0.9 times F2 length; face, frons, vertex, and clypeus smooth, setiferous; gena, temple, and eye smooth, glabrous; mandible bidentate, setiferous; occipital carina prominent, complete, meeting hypostomal carina.

Mesosoma. Mesoscutum smooth, setiferous; small, shallow groove present centrally on median mesonotal lobe extending to anterior portion of scutellum; notauli present anteriolaterally, dorsally effaced; scutellum smooth, setiferous; mesopleuron glabrous, smooth, and shining with short, deep sternalus; epicnemial carina present; propodeum smooth to rugose, setiferous with prominent pentagonal areola and complex carination.

Wings. Forewing pterostigma length 5.0 times width; wings hyaline, setose, venation light brown; 3RS tubular proximally, spectral to absent distally, not reaching wing margin; 3RS length subequal to R1 and 1.5 times length of 2M; (RS+M)b tubular; r and 1RS of equal length. Hindwing hyaline, setose, venation light brown; C+Sc+R, SC+R, 1M, M+Cu, and R1 tubular; all other veins absent.

Metasoma. Tergite 1 (T1) 1.5 times wider than long, heavily sclerotized, aciculo-rugose centrally and aciculate to smooth laterally; T2 and T3 fused, 1.1 times as long as wide; T2 with light aciculae medially, smooth and shining laterally; aciculae ending just beyond T2-T3 fusion; T3 smooth and shining; ovipositor sheath short, setose with a few especially elongate setae, bisected dorso-ventrally and exserted from the ventral side of the metasoma.

Coloration. Head dark brown, scape, pedicel, and proximal half of F1 yellowish, remaining segments dark brown; mesosoma dark brown; metasomal T1 and T2 orangish-brown, T3 and remaining tergites dark brown; legs yellowish and distal one-third to one-half of hind tibia dark brown.

Male. As for female except sexual differences.

Etymology: The species epithet "*erwini*" is a patronym honoring the late Dr. Terry L. Erwin of the Smithsonian Institution National Museum of Natural History and his comment in Rice (2015) about the paucity of active hymenopterists, as well as our admiration for his biodiversity research in the Amazonian rainforest.

Distribution. U.S.A., South Dakota.

Biology. The wasp specimens cited were reared from licorice pod samples that overwintered within a clear plastic sack in a room temperature lab. A list of the insects associated with licorice in South Dakota is presented in Table 2 with other recorded associates in Table 3.

***Hystrichophora ochreicostana* (Walsingham)** (Figure 13)

Hystrichophora ochreicostana (Walsingham) (Lepidoptera: Tortricidae: Enarmoniini) was originally described as *Semasia ochreicostana* (Walshingham 1884) based on two females from Montana. It is a native species now known from the prairie and intermountain regions of Alberta, Manitoba, Ontario, Saskatchewan, Colorado, Iowa, Kansas, Kentucky, Minnesota, Missouri, Montana, Nebraska, North Dakota, South Dakota, Texas, Utah, and Wyoming (Heinrich 1923; Metzler *et al.* 2005; Gilligan *et al.* 2008; BMNA 2023; NAMPG 2023). Gillette (1897) reported a specimen from leaves of boxelder (*Acer negundo* L.), probably as a resting site, otherwise no reference seen provided host information.

Three adults were reared from a bulk sample of pods of *Gly. lepidota* and are labelled as SOUTH DAKOTA: Sanborn Co., 2.5 mi. W. of Woonsocket, along Hwy 34, 44.061N, -98.305W, elev. 1310 ft., coll. 27 July 2023, P.J. Johnson / ex *Glycyrrhiza lepidota* pods; **new host record**. The first specimen emerged on 4 August 2023, the second on 5 August, and the third on 7 August. No leaves or stems of *Gly. lepidota,* or other plant material were part of the sample.

Figure 13. ***Hystrichophora ochreicostana*** **female reared from licorice.**

Hemiptera
Alydidae
***Alydus eurinus* (Say)**

This common and widespread broad-headed bug (Fracker 1918) feeds on a variety of legumes and other plants and can be a pest on soybean (*Glycine max* (L.) Merrill), lima bean (*Phaseolus lunatus* L.), and cowpea (*Vigna unguiculata* (L.) Walp. (Fabales: Fabaceae) (Underhill 1943), while Schuh and Slater (1995) generalized that all species in the subfamily Alydinae feed on legumes. Blatchley (1926) noted that this species and *A. pilosulus* often co-occur on the same hosts. Live specimens for this study were collected in the summer of 2023 from the Felt Farm, Innovation Campus, Lake Poinsett, Rasmussen, and Summit Lake GPA locations as given above. We observed feeding consists of probing a mature pod that is still green, piercing the valve integument, as for *Al. pilosulus* (Figure 14), and feeding on an encountered green seed (Figure 15). In licorice we found that the feeding on green seeds results in a highly variable 20–100% mortality per pod. Damage to valve tissues results in bubbles and patches of amber colored resin on outside and inside surfaces of a pierced valve at the point of penetration. *Alydus eurinus* was listed by Parshley (1922) and Harris (1937) and confirmed from eastern and western South Dakota by Swanson (2018), though apparently licorice is a **new host record** (Underhill 1943; Yonke and Medler 1968; Schaefer 1980; Schaefer and Mitchell 1983; Froeschner 1988). Fracker (1918) reported the species from "Dakota" based on Uhler (1878, p. 384), but that may be an error as the latter author reported *Neides muticus* (Say) from that vague location lower on the page.

***Alydus pilosulus* (Herrich-Schaeffer)** Figure 14

The general biology of this species was treated by Underhill (1943) and Wilkinson and Daugherty (1967). This common and widespread broad-headed bug feeds on green seeds and often co-occurs with the similar appearing *Al. eurinus* on

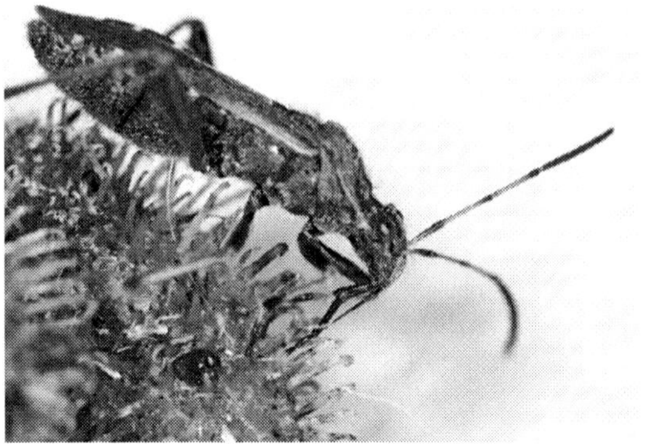

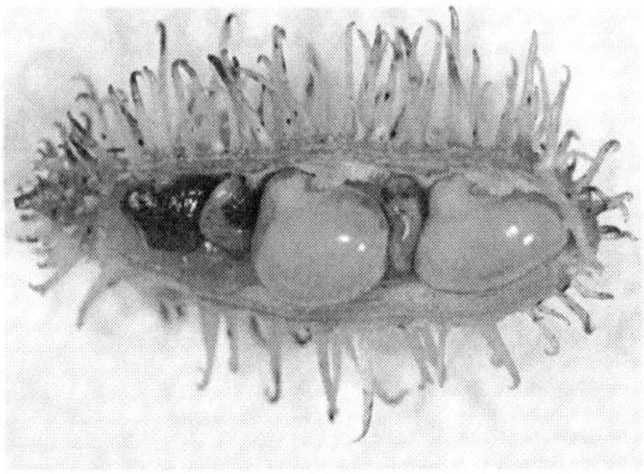

Figure 14. (left) *Alydus pilosulus* piercing the immature valve of a licorice pod. Note proboscis from front of head. Figure 15. (right) Feeding damage to immature seeds of licorice by *Al. eurinus* or *Al. pilosulus*. Damaged seeds are, left to right, number 1, 2 and 4. Seeds 3 and 5 are normal undamaged seeds.

licorice pods in our region and has been reported from western South Dakota by Parshley (1922) and Harris (1937). Live specimens for this study were collected during the summers of 2022 and 2023 from the Felt Farm, Innovation Campus, Rasmussen, and Summit Lake GPA locations as given above, usually along with *Aly. eurinus* and *Eus. variolarius*. Combined feeding by one or both bugs, and sometimes with the stink bug *Eus. variolarius* (see below) results in 20-100% seed mortality per pod. Like *Al. eurinus*, observed feeding consists of probing through a valve of a mature green pod and feeding on an encountered seed (Figure 15) that turns brown and shrivels, as noted on other legumes by Underhill (1943). Damage to valve tissues results in bubbles and patches of amber colored resin on outside and inside surfaces of the valve at the point of penetration. This species apparently has not been reported previously from *Gly. lepidota* (Wilkinson and Daugherty 1967; Yonke and Medler 1968; Schaefer 1980; Schaefer and Mitchell 1983), hence provides a **new host record**. Schaefer and Panizzi (2000) regarded this species as a minor pest on soybeans and other legumes.

Pentatomidae
***Euschistus variolarius* (Palisot de Beauvois)** (Figure 16)

A widespread species across North America (McPherson 1982; Schaefer and Panizzi 2000; Rider 2012; Rider and Swanson 2021) had been reported early from South Dakota by Parshley (1922) and Harris (1937). Live specimens for this study were collected during the summer of 2023 from the Felt Farm, Innovation Campus, Rasmussen, and Summit Lake GPA locations as given above, usually along with the two *Alydus* spp. Panizzi (1997) stated that phytophagous stink bugs such as *Eus. variolarius* are polyphagous, opportunistic, apparently lacking host preference, and can be a serious pest of cultivated plants, including soybean and other legumes. The feeding on licorice by this bug results in damage indistinguishable from that of the *Alydus* species above (Figure 15). Pezzini *et al.* (2019) noted that *Eus. variolarius* was one of the native stink bugs that are demonstrating increases in population and abundance in the north-central United States, commonly feeding on soybean pods. This finding bodes poorly for licorice that grows in proximity to soybean: we found over 23% of licorice seed loss and reproductive capacity (Table 1) was due to *Eus. variolarius* and the two *Alydus* species noted above. It seems that the growing of soybean presents an ecological threat to licorice by its support of legume feeding insects.

Euschistus variolarius overwinters as an adult and is univoltine (Munyaneza and McPherson 1994). As a phytophagous bug, feeding is by injecting salivary enzymes and removing liquified contents; feeding channels marked by resin are remnants of activity. Observed damage to licorice valves and seeds is similar to that described by Panizzi (1997) to soybean and to corn (*Zea mays* L.) (Venugopal *et al.* 2014). Parish (1934) listed a number of host plants for *Eus. variolarius* which did not include licorice. Reviewing the literature indicates little attention to native hosts for this species. This report of feeding on licorice by *Eus. variolarius* seems to provide a **new host record** (McPherson and McPherson 2000).

Figure 16. *Euschistus variolarius* on seed pod cluster of licorice.

Additional Perspectives

American licorice has a long botanical, vegetational, and cultural history across the continent. Its structural architecture and biology causes pause as to why it was essentially ignored by most biologists. Licorice is unique among the native legumes of the region: it is the largest, and it spreads primarily by rhizomes in often marginal, transitional, or ecotonal habitats, and has epizoochoric seed distribution. A 12% occurrence of lethal albinism was observed from seeds produced in natural populations, due putatively to homozygosity of recessive alleles leading to chlorophyll deficiency in seedlings (Wynia *et al.* 1981; Boe and Wynia 1985). Whether licorice is facultatively xenogamous remains to be thoroughly tested.

In classic prairie studies and summaries, Weaver and Hansen (1941a–b) and Weaver (1968) regarded licorice as one of 12 important and characteristic prairie legumes and one of 32 forb species in prairies that respond negatively or decrease in presence due to grazing. The relative lack of attention, however, provides many opportunities for investigations on the plants' biology and its insect and other associates. Given the ecological and environmental changes in North American over the past 250 years, licorice can be a model organism for examining abiotic dynamics in changing ecosystems, as well as its community ecology with multiple trophic levels of associated animals.

Three major ecological and biodiversity issues that are suggested in our study that we feel are worth further attention: 1) if toxicants occur in licorice seed, as in some other legume seeds (Janzen 1969) and dissuade seed predators, there may be interesting secondary plant chemicals involved that act in possible combination with indehiscence of the pod to eliminate or reduce bruchid larval feeding; 2) habitat fragmentation may have a direct and multi-layered impact on the viability of licorice populations and the insects associated with the plant; and 3) the role of adventitious chalcidoid wasp parasitoids that are invasive in natural communities and induce replacement of and at least local extinction of native seed eaters and their native parasitoids.

Issues two and three are of particular concern as those are anthropogenic. Janzen and Hallwachs (2019) was among the first to discuss the impact of anthropogenic activities on the decline of insect occurrences, a concept that is increasingly studied (Wagner *et al.* 2021) and is well-displayed by native plant and insect distributions in the remnant and disturbed prairies of the central United States and is exemplified by licorice. Conversion or disfigurement of landscapes from native communities in the northern Great Plains has received considerable, if not tardy, interest in its impacts (*e.g.*, Wright and Wimberly 2013; Johnston 2014; Wick *et al.* 2016; Wimberly *et al.* 2017; Hendrickson *et al.* 2019). Regional agriculturalization through conversion of native communities, inadvertent introduction of invasive weeds, high density or frequent grazing, herbicides, mowing, haying of meadows, roadsides, field margins and degradation of environmental characteristics through treeing and other "development" induces habitat fragmentation and eliminates populations directly. The fragmentation of livable habitats and scattering of viable populations opens potential for studying isolated island-like biogeographic patterns, problems of source area migrations of seed-feeders and their parasitoids, habitat utilization patterns, and the negative anthropogenic impacts on native communities and guilds such as pollinators.

Licorice is considered a weed in some states, such as South Dakota (SDDA 2009), though its occurrence on generally marginal sites for crops would restrict any minimal impact to pastures. We have observed significant impact on established populations by mowing and haying and presumably roadside herbicide use that directly impacts plant viability through removal of shoots and flowers, that reduces energy transfer to rhizomes, pollinator visitations and physically prevents epizoochoric redistribution and removal of seed guilds. These deleterious impacts may cause local extinctions of populations if repeated annually. Without disturbances, pod set is generally high with >50% of flowers producing pods and nearly 30% of seed production being viable in natural populations (Table 1; AB pers. observ.). Details on the full breeding system of licorice have not been studied and could prove interesting.

In the northern Great Plains the domination of grassland, roadsides, untilled fields, and pastures by invasive smooth bromegrass (*Bromus inermis* Leyss.),

Kentucky bluegrass (*Poa pratensis* L.) and reed canary grass (*Phalaris arundinacea* L.) (*e.g.*, DeKeyser *et al.* 2015; Hendrickson and Lund 2010; Hendrickson *et al.* 2020) alone have severely stressed native communities toward effective local extinction of native species. In contrast, some observations on new populations (*e.g.*, Innovation Campus, Felt Farm, Oak Lake Field Station) and large established ones (*e.g.*, Rasmussen GPA, 6-Mile Creek) suggest licorice rhizomatous growth and ability to fix nitrogen in nutrient limited, marginal soils allows it to tolerate Kentucky bluegrass, and to some degree smooth brome competition if it preceded the grass on the site. Some of our sampling sites are patches of many thousands of shoots that, if the plants are not hayed, grazed, sprayed or otherwise frequently disturbed, exist with Kentucky bluegrass and are evidently stable and increasing in area. We suggest that licorice may actually be a good potential species for vegetative regeneration of agriculturally marginal sites that are not subject to heavy grazing or other frequent community disturbances.

Now that we are finding *M. bruchophagus* in *Ac. aureolus,* and as *B. grisselli* and its reported native parasitoids seem lacking, a working hypothesis is that the invasive and opportunistic parasitoid *M. bruchophagus* has moved to *Ac. aureolus* while seemingly overwhelming the *B. grisselli* population and its previously reported native parasitoids. This wasp (as *Eutelus bruchophagi*) was first reported in North America as a parasitoid of the alfalfa (*Medicago sativa* L.) chalcid *Bruchophagus roddi*, then later reported in *B. platypterus*, *B. gibbus* and *B. grisselli*, as well as various weevils and bruchids (Urbahns 1914, 1919, 1920). This appears to be an example of parasitoid drift, where an introduced parasitoid moves from its original host to unintended native hosts (Follett *et al.* 2000). Alternatively, it cannot be discounted that some of the variability in seed-eater and parasitoid populations may be natural ebb and flows of fluctuations that have yet to be detected. Also, the impact of *Gr. interstinctana* causing considerable seed loss in our samples (Table 1) in combination with predation by *Alydus* spp, *Euschistus*, and *Ac. aureolus* and then adding the invasive *M. bruchophagus* parasitoid may provide seed competition that overwhelms and depresses *B. grisselli* populations to the point of undetectability.

The introduced parasitoids and the observations given above, especially with regard to *B. grisselli*, may provide another example of the impact of introducing exotic biological control hopeful-helpfuls toward nontarget native species (*e.g.*, Howarth 1983; Follett *et al.* 2000). *Dinarmus acutus,* found now in the licorice seed-eater parasitoid guild, was introduced to Oregon for the control of the hairy vetch (*Vicia villosa* Roth) bruchid (*Bruchus brachialis* Fahraeus) (Ritcher 1966). Leong and Dickason (1975) then noted that *D. acutus* along with *Baryscapus bruchivorus* (Gahan) (as *Tetrastichus bruchivorus*) introduced in 1965 and *Eup. vesicularis* earlier in 1915, determined "that the parasite [*D. acutus*] was not an effective vetch bruchid population factor under Oregon conditions", and neither were *Ba. bruchivorus* or *Eup. vesicularis.* Still, each species subsequently became adapted to other bruchid beetles and numerous other insects and spread across the continent and invaded native parasitoid guilds. Another introduced parasitoid, *Eupelmus vesicularis*, alone has seemingly an expansive taste for many kinds of insects (Burks 1979). Thus, the occurrence of a diverse community of seed predators, their native and adventitious parasitoids, and other insects associated with licorice suggests a greater ecological role for licorice in natural and renovated communities than previously assumed.

ACKNOWLEDGEMENTS

Our gratitude is extended to Owen McElroy, South Dakota Department of Game, Fish and Parks, for assistance in locating sampling sites in northeastern South Dakota, and Mike McKernan, Hull/McKernan Ranch, Twin Brooks, SD, for access to sites on ranch property, advice on grazing plans, and area history. Thomas Stuart is thanked for permission to use his image of an adult *Gr. interstinctana* that is also posted on Bugguide.net. Mark Gabel, Spearfish, SD, generously provided samples from across the state. Mary and Tom Bodensteiner, Rapid City, kindly provided samples from the Black Hills and logistical support for fieldwork. The Six Mile Creek Golf Course, White, South Dakota, is thanked for permitting access to streamside portions of their property, as are the South Dakota Crop Improvement Association and the Felt Family Farm through the South Dakota State University Agricultural Experiment Station for access to research farms. Louis Hesler, USDA-ARS, Brookings, SD, Brian Patrick, Dakota Wesleyan University, Mitchell, SD, and David A. Rider, North Dakota State University, Fargo, ND, kindly provided presubmission reviews and beneficial comments and corrections on this manuscript. This research was supported in part by the USDA-NIFA through the South Dakota Agricultural Experiment Station at South Dakota State University.

LITERATURE CITED

Albers, P.C., C. Berndt, E. Brown, Y. Kelly and V. Kittelson. 2003. The Home of the Bison: An ethnographic and ethnohistorical study of traditional cultural affiliations to Wind Cave National Park. U.S. National Park Service Publications and Papers 158. xlviii + 440 p. (Available at http://digitalcommons.unl.edu/natlpark/158)

Allen, H.W. 1962. Parasites of the Oriental fruit moth. United States Department of Agriculture Technical Bulletin No. 1265, ii + 139 p.

Allen, O.N. and E.K. Allen. 1980. The Leguminosae, A Source Book of Characteristics, Uses and Nodulation. University of Wisconsin Press, Madison, WI. lxiv + 812 p.

Andersen, D.C. and S.M. Nelson. 2014. Effects of soil temperature and depth to ground water on first-year growth of a dryland riparian phreatophyte, *Glycyrrhiza lepidota* (American licorice). Southwestern Naturalist 59(1): 56–65.

Asquith, A. 1990. Taxonomy and evolution of the orthotyline genus *Lopidea* (Heteroptera: Miridae). Unpubl. PhD thesis, Oregon State University, Corvallis. xvii + 404 p.

Baker, C.F. 1895. Biological notes on some Colorado Coleoptera. Entomological News 6: 27–29.

Barghi, N. and S. Siljak-Yakovlev. 1990. Karyological study in three species of *Glycyrrhiza* genus (*G. glabra*, *G. lepidota* and *G. echinata*). Caryologia 43(3–4): 223–234.

Beal, W.J. 1897 [1896]. Grasses of North America, second edition. In two volumes. Vol. 1: Chapters on the physiology, composition, selection, improving, and cultivation of grasses; management of grassland; also chapters on clovers, injurious insects, and fungi. Henry Holt and Company, New York. xiii + 457 p. [title page dated 1896, preface dated 1897]

Berthoud, E.L. 1892. A peculiar case of plant dissemination. Botanical Gazette 17(10): 321–326.

Bjugstad, A.J. and W. C. Whitman. 1984. Perennial forbs for wildlife habitat restoration on mined lands in the northern Great Plains [pp. 257–271]. *In*: Proceedings of the Annual Conference of the Western Association of Fish and Wildlife Agencies and the Western Division American Fisheries Society. Western Association of Fish and Wildlife Agencies, July 18–22, 1982. Las Vegas, Nevada. 622 p.

Bjugstad, A.J. and W. C. Whitman. 1989. Promising native forbs for seeding on mine spoils [pp. 255–262]. *In*: Wolters, Gale L.; Aldon, Earl R.; Donart, Gary B. Proceedings of the Conference Reclamation, a Global Perspective: held in Calgary, Alberta, Canada, August 27-31, 1989. Report RRTAC 89-2. [Edmonton]: Alberta Land Conservation and Reclamation Council, Reclamation Research Technical Advisory Committee: 255–262.

Blatchley, W.S. 1926. Heteroptera or True Bugs of Eastern North America with Especial Reference to the Faunas of Indiana and Florida. The Nature Publishing Company, Indianapolis, IN. 1116 p.

BMNA (Butterflies and Moths of North America. 2023. *Hystrichophora ochreicostana* (Walsingham). https://www.butterfliesandmoths.org/species/hystrichophora-ochreicostana

Boe, A., I. Akundabweni and D. Walgenbach. 1981. Seed predation by *Bruchus fraterculus* Horn in American licorice (*Glycyrrhiza lepidota* Pursh) populations from North and South Dakota. Proceedings of the South Dakota Academy of Science 60: 172.

Boe, A. and P.J. Johnson. 2008. Seed predation by *Acanthoscelides submuticus* and *A. pallidipennis* (Coleoptera: Bruchidae) in false indigo in the northern Great Plains. Proceedings of the South Dakota Academy of Science 87: 223–228.

Boe, A. and P.J. Johnson 2015. First report of seed predators of Platte milkvetch, *Astragalus plattensis* Nutt, ex T.&G. Proceedings of the South Dakota Academy of Science 94: 297–301.

Boe, A., and P.J. Johnson. 2016. Reproductive biology of American licorice (*Glycyrrhiza lepidota* Pursh) in western South Dakota with new seed insect associations. Proceedings of the South Dakota Academy of Science 95: 91–103.

Boe, A., P.J. Johnson and A.B. Martens. 2019. Seed set and seed-insect interactions in natural and cultivated populations of purple prairie clover. Proceedings of the South Dakota Academy of Sciences 98: 105–118.

Boe, A. and B. McDaniel. 1989. Emergence patterns of chalcids associated with alfalfa seed [pp. 9–10]. Proceedings of the Twenty-First Central Alfalfa Improvement Conference, July 17–20, 1989, University of Wisconsin, Madison, Wisconsin. [2] + iv + 53 p.

Boe, A. and B. McDaniel. 1990. Population biology of the alfalfa seed chalcid and its chalcidoid parasites, p. 23. Report of the Thirty-Second North American Alfalfa Improvement Conference, August 19–24, 1990, Pasco, Washington. [9] + 153 p.

Boe, A. and B. McDaniel. 1991. New host and distribution records for *Uscana semifumipennis* (Hymenoptera: Trichogrammatidae), an egg parasite of bruchid beetles. Prairie Naturalist 23: 108.

Boe, A., B. McDaniel and K. Robbins. 1988. Patterns of American licorice seed predation by *Acanthoscelides aureolus* (Horn) (Coleoptera: Bruchidae) in South Dakota. Journal of Range Management 41(4): 342–345.

Boe, A., K. Robbins, N. Foster and B. McDaniel. 1990. Reproductive biology of *Glycyrrhiza lepidota* Pursh. Proceedings of the South Dakota Academy of Science 69: 172–173 [abstract].

Boe, A., and R. Wynia. 1985. Seed predation, seedling emergence, and rhizome characteristics of American licorice. Journal of Range Management 38(5): 400–402.

Britton, N.L. and A. Brown. 1913. An Illustrated Flora of the Northern United States, Canada and the British Possessions, from Newfoundland to the Parallel of the Southern Boundary of Virginia, and from the Atlantic Ocean Westward to the 102D Meridian. Second Edition – Revised and Enlarged, Vol. II. Amaranthaceae to Loganiaceae. Amaranth to Polypremum. Charles Scribner's Sons, New York. iv + 735 p.

BugGuide 2023. *Hystrichophora ochreicostana* – Hodges#3396. https://bugguide.net/node/view/800395/data.

Brown, J.W. 2022. A review of host plants for the tortricid tribe Grapholitini, with a synopsis of host utilization by genus (Lepidoptera: Tortricidae). Insecta Mundi 0944: 1–75.

Bråthen, K.A., F.I. Pugnaire and R.D. Bardgett. 2021. The paradox of forbs in grasslands and the legacy of the mammoth steppe. Frontiers in Ecology and the Environment 19(10): 584–592.

Burks, B.D. 1979. Eurytomidae [pp. 835–860]. *In*: K.V. Krombein, P.D. Hurd, Jr., D.R. Smith and B.D. Burks, volume 1, Catalog of Hymenoptera in America North of Mexico. Smithsonian Institution Press, Washington, D.C. xvi + 1–1198 p.

CalFlora. 2023. *Glycyrrhiza glabra* L., Cultivated licorice. https://www.calflora.org/app/taxon?crn=3932.

Çetin, Ö., A. Duran, E. Martin and M. Küçüködük. 2015. Karyological studies in some *Glycyrrhiza* (Fabaceae) taxa from Turkey. Caryologia 68(3): 254–264.

Clemens, B. 1861. Contributions to American Lepidopterology. – No. 6. Proceedings of the Academy of Natural Sciences of Philadelphia (1860) 12: 345–362.

Clemens, B. 1865. North American Micro-Lepidoptera. Proceedings of the Entomological Society of Philadelphia 5: 133–147.

Comstock, J.H. 1881. Report of the Entomologist [pp. 235–373, 24 pl.]. Annual Report of the Commissioner of Agriculture for the Year 1880. Washington, D.C. 672 p.

Cushman, R.A. 1911. Notes on the host plants and parasites of some North American Bruchidae. Journal of Economic Entomology 4(6): 489–510.

DeKeyser, E.S., L.A. Dennhardt and J. Hendrickson. 2015. Kentucky bluegrass (*Poa pratensis*) invasion in the Northern Great Plains: A story of rapid dominance in an endangered ecosystem. Invasive Plant Science and Management 8: 255–261.

Ding, Y., E. Brand, W. Wang, and Z. Zhao. 2022. Licorice: Resources, applications in ancient and modern times. Journal of Ethnopharmacology 298, https://doi.org/10.1016/j.jep.2022.115594.

Dodd, J.D., and R.T. Coupland. 1966. Vegetation of saline areas in Saskatchewan. Ecology 47: 958–968.

Duan, L., A.J. Harris, C. Su, Z.-R. Zhang, E. Arslan, K. Ertuğrul, P.K. Loc, H. Hayashi, J. Wen and H.-F. Chen. 2020. Chloroplast phylogenomics reveals the intercontinental biogeographic history of the liquorice genus (Leguminosae: *Glycyrrhiza*). Frontiers in Plant Science 11, doi: 10.3389/fpls.2020.00793.

Duan, L., L.-N. Han, B.-B. Liu, A. Leostrin, A.J. Harris, L. Wang, E. Arslan, K. Ertufğrul, M. Knyazev, E. Hantemirova, J. Wen and H.-F. Chen. 2022. Species delimitation of the liquorice tribe (Leguminosae: Glycyrrhizeae) based on phylogenomic and machine learning analyses. Journal of Systematics and Evolution, https://doi.org/10.1111/jse.12902.

Duke, J.A. 1981. Handbook of Legumes of World Economic Importance. Plenum Press, New York. xi + 345.

Ellis, C.R. and F.L.O. Nang'ayo. 1992. The biology and control of the trefoil seed chalcid, *Bruchophagus platypterus* (Walker) (Hymenoptera: Eurytomidae), in Ontario. Proceedings of the Entomological Society of Ontario 123: 111–122.

Esser, L.L. 1994. *Glycyrrhiza lepidota*. *In*: Fire Effects Information System (FEIS) [Online]. U.S. Department of Agriculture, Forest Service, Rocky Mountain Research Station, Fire Sciences Laboratory (Producer). Available at: https://www.fs.usda.gov/database/feis/plants/forb/glylep/all.html [downloaded 24 January 2024].

Eyheralde, P.G. 2015. Bison-mediated seed dispersal in a tallgrass prairie reconstruction. PhD dissertation, Iowa State University, Ames, Iowa. iii + 154 p.

Fernald, M.L. 1950. Gray's Manual of Botany. Eighth (Centennial Edition) – Illustrated. A Handbook of the Flowering Plants and Ferns of the Central and Northeastern United States and Adjacent Canada. Dioscorides Press, Portland, Oregon. lxiv + 1632 p.

Follett, P.A., J. Duan, R.H. Messing and V.P. Jones. 2000. Parasitoid drift after biological control introductions: Re-examining Pandora's box. American Entomologist 46(2): 82–94.

Folsom, J.W. 1909. The insect pests of clover and alfalfa. University of Illinois Agricultural Experiment Station Bulletin No. 134, Urbana, Illinois.

Fracker, S.B. 1918. The Alydinæ of the United States. Annals of the Entomological Society of America 11: 255–280, pl. 24–25.

Froeschner, R.C. 1988. Family Alydidae Amyot and Serville, 1843. The Broad-Headed Bugs. [pp. 4–11]. *In*: T.J. Henry and R.C. Froeschner (eds.), Catalog of the Heteroptera, or TrueBugs, of Canada and the Continental United States. St. Lucie Press, Boca Raton, Florida. xix + 978 p.

Gabel, M., D. Ode, G. Kostel and C. Card. 2010. Early vascular plant collections from eastern Wyoming and western South Dakota from 1800–1940. Proceedings of the South Dakota Academy of Science 89: 69–76.

Gabel, M., D. Ode and G. Kostel. 2015. A history of botanical exploration in South Dakota. Proceedings of the South Dakota Academy of Science 94: 69–83.

Gillette, C.P. 1897. Colorado Lepidoptera. Colorado Agricultural Experiment Station Bulletin No. 43, 31 p.

Gilligan, T.M. and M.E. Epstein. 2014. Tortricids of Agricultural Importance. https://idtools.org/id/leps/tortai/index.html (last accessed 6 August 2023).

Gilligan, T.M., D.J. Wright and L.D. Gibson. 2008. Olethreutine Moths of the Midwestern United States. An Identification Guide. Bulletin of the Ohio Biological Survey (New Series), vol. 16, No. 2, vii + 334 p.

Gossard, H.A. 1892. The clover-seed caterpillar [pp. 571–589]. *In* Report of experiments and studies in entomology. Iowa Agricultural Experiment Station Bulletin No. 19: 535–636.

GPFA (Great Plains Flora Association). 1986. Flora of the Great Plains. University Press of Kansas, Lawrence. vii + 1402 p.

Grissell, E.E. 2005. A review of North American species of *Microdontomerus* Crawford (Torymidae: Hymenoptera). Journal of Hymenoptera Research 14(1): 22–65.

Grote, A.R. 1873. Contributions to a knowledge of North American moths. Bulletin of the Buffalo Society of Natural Sciences 1: 73–94.

Harris, H.M. 1937. Contributions to the South Dakota list of Hemiptera. Iowa State College Journal of Science 11: 169–176.

Hayashi, H. E. Miwa and K. Inoue. 2005. Phylogenetic relationship of *Glycyrrhiza lepidota*, American Licorice, in genus *Glycyrrhiza* based on *rbc*L sequences and chemical constituents. Biological and Pharmaceutical Bulletin 28(1): 161–164.

Heinrich, C. 1923. Revision of the North American moths of the subfamily Eucosminae of the family Olethreutidae. United States National Museum Bulletin 123, Washington, D.C. iv + 298 p. + 59 pl.

Heinrich, C. 1926. Revision of the North American moths of the subfamilies Laspleyresiinae and Oleuthreutinae. United States National Museum Bulletin 132, Washington, D.C. iii + 216 p. + 76 pl.

Hendrickson, J.R. and C. Lund. 2010. Plant community and target species affect responses to restoration strategies. Rangeland Ecology & Management 63: 435–442.

Hendrickson, J.R., S.L. Kronberg and E.J. Scholljegerdes. 2020. Can targeted grazing reduce abundance of invasive perennial grass (Kentucky bluegrass) on native mixed-gras prairie? Rangeland Ecology & Management 73: 547–551.

Hendrickson, J.R., K.K. Sedivec, D. Toledo and J. Printz. 2019. Challenges facing grasslands in the northern Great Plains and North Central Region. Rangelands 41(1): 23–29.

Hetz, M. and C.D. Johnson. 1988. Hymenopterous parasites of some bruchid beetles of North and Central America. Journal of Stored Products Research 24(3): 131–143.

Hobbs, G.A. 1966. Ecology of species of *Bombus* Latr. (Hymenoptera: Apidae) in southern Alberta. V. Subgenus *Subterraneobombus* Vogt. The Canadian Entomologist 98: 288–294.

Hoffman, G.R. and D.J. Rogers. 1966. A bibliography of vegetation studies in the state of South Dakota. Proceedings of the South Dakota Academy of Science 45: 164–169.

Howarth, F.G. 1983. Classical biological control: panacea or Pandora's box? Proceedings of the Hawaiian Entomological Society 24: 239–244.

Huan, C., Y. Xu, W. Zhang, T. Guo, H. Pan and S. Gao. 2021. Research progress on the antiviral activity of glycyrrhizin and its derivatives in liquorice. Frontiers in Pharmacology 12: 680674. 10 p. Doi: 10.3389/fphar.2021.680674

Hurd, P.D., Jr. 1979. Apoidea [pp. 1741–2209]. *In*: K.V. Krombein, P.D. Hurd, Jr., D.R. Smith and B.D. Burks, volume 2, Catalog of Hymenoptera in America North of Mexico. Smithsonian Institution Press, Washington, D.C. xvi + 1199–2209 p.

Janzen, D.H. 1969. Seed-eaters versus seed size, number, toxicity and dispersal. Evolution 23(1): 1–27.

Janzen, D.H. 1974. Deflowering of Central America. Natural History 83: 48–53.

Janzen, D.H. and W. Hallwachs. 2019. Where might be many tropical insects? Biological Conservation 233: 102–108.

Johnson, C.D. 1970. Biosystematics of the Arizona, California, and Oregon species of the seed beetle genus *Acanthoscelides* Schilsky (Coleoptera: Bruchidae). University of California Publications in Entomology No. 59, vi + 113 p.

Johnson, C.D. 1981. Interactions between bruchid (Coleoptera) feeding guilds and behavioral patterns of pods of the Leguminosae. Environmental Entomology 10: 249–253.

Johnson, J.R. and G.E. Larson. 1999. Grassland Plants of South Dakota and the Northern GreatPlains. South Dakota Agricultural Experiment Station B 566 (rev.), South Dakota State University, Brookings, South Dakota. 288 p.

Johnson, J.R. and J.T. Nichols 1970. Plants of South Dakota Grasslands: A photographic study. South Dakota Agricultural Experiment Station Bulletin 566, South Dakota State University, Brookings, South Dakota. 163 p.

Johnston, C.A. 2014. Agricultural expansion: land use shell game in the U.S. northern plains. Landscape Ecology 29: 81–95.

Kartesz, J.T. 2015. The Biota of North America Program (BONAP). 2015. North American Plant Atlas. (http://bonap.net/napa). Chapel Hill, N.C. [maps generated from Kartesz, J.T. 2015. Floristic Synthesis of North America, Version 1.0. Biota of North America Program (BONAP).

Kindscher, K. 1987. Edible wild plants of the prairie: An ethnobotanical guide. University Press of Kansas, Lawrence. 288 p.

Kindscher, K. 1992. Medicinal wild plants of the prairie: An ethnobotanical guide. University Press of Kansas, Lawrence. 336 p.

Kingsolver, J.M. 2004a. Handbook of the Bruchidae of the United States and Canada (Insecta, Coleoptera). Volume 1. United States Department of Agriculture, Agricultural Research Service Technical Bulletin Number 1912, xi + 324 p.

Kingsolver, J.M. 2004b. Handbook of the Bruchidae of the United States and Canada (Insecta, Coleoptera). Volume 2 (Illustrations). United States Department of Agriculture, Agricultural Research Service Technical Bulletin Number 1912, ii + 198 p.

Kirsch, L.M. and A.D. Kruse. 1973. Prairie fires and wildlife [pp. 289–304]. Fire Ecology Conference Proceedings, vol. 12, Tall Timbers Research Station, Tallahassee, FL. 488 p.

Krombein, K.V., P.D. Hurd, Jr., D.R. Smith and B.D. Burks. 1979. Catalog of Hymenoptera in America North of Mexico (3 volumes). Smithsonian Institution Press, Washington, D.C. xvi + 1–1198 p.

Kulbaba, M.W., J.C. Tardif and R.J. Staniforth. 2009. Morphological and ecological relationships between burrs and furs. The American Midland Naturalist 161(2): 380–391.

Lam, W.H.Y., J. Rota and J.W. Brown. 2011. A preliminary list of the leaf-roller moths (Lepidoptera: Tortricidae) of Virginia. Banisteria 38: 3–37.

Larson, G.E. and J.R. Johnson. 1999. Plants of the Black Hills and Bear Lodge Mountains. South Dakota Agricultural Experiment Station B732, South Dakota State University, Brookings, South Dakota. 608 p.

Leong, K.L. and E.A. Dickason 1975. Biology of *Dinarmus acutus*, a chalcidoid parasite of the vetch bruchid. Annals of the Entomological Society of America 68(6): 943–948.

Lewis, M., Clark, W., and Members of the Corps of Discovery. 2002. Wednesday, May 8th, 1805. *In*: G. Moulton (ed.), *The Journals of the Lewis and Clark Expedition*. Lincoln: University of Nebraska Press. Retrieved 16 December 2023, from the University of Nebraska Press / University of Nebraska-Lincoln Libraries-Electronic Text Center, The Journals of the Lewis and Clark Expedition web site: http://lewisandclarkjournals.unl.edu/journals.php?id=1806-09-04.

Lintner, J.A. 1896. *Grapholitha Interstinctana* (Clemens), The Clover-seed Caterpillar [pp. 152–156]. *In*: Eleventh Report on the Injurious and Other Insects of the State of New York for the Year 1895. 49th Report on the New York State Museum. University of the State of New York, Albany. pp. 89–325.

Mackay, M.R. 1959. Larvae of the North American Olethreutidae. The Canadian Entomologist, Supplement 10, 338 p.

Manglitz, G.R. 1985. Insects and Related Pests [pp. 269–294]. *In* N.L. Tayler (ed.), Clover Science and Technology. American Society of Agronomy, Inc., Crop Science Society of America, Inc., Soil Science Society of America, Inc., Publishers, Madison, Wisconsin. xx + 616 p.

Marsh, P.M. and R.W. Carlson. 1979. Ichneumonoidea [pp. 143–742]. *In*: K.V. Krombein, P.D. Hurd, Jr., D.R. Smith and B.D. Burks, volume 1, Catalog of Hymenoptera in America North of Mexico. Smithsonian Institution Press, Washington, D.C. xvi + 1198 p.

Martens, A.P. 2019. A review of the genus *Acrisis* Foerster (Hymenoptera: Braconidae) with a new species from American licorice and a survey of aphid parasitoids of soybean fields ineastern South Dakota. Master of Science thesis, South Dakota State University, Brookings. 41 p.

Matteson, R.L. 1933. The development of the floral organs of *Glycyrrhiza lepidota*. M.S. thesis, South Dakota State College, Brookings, South Dakota. 21 p.

McDaniel, B. and A. Boe. 1991a. Life history studies, host records, and morphological description of genitalia of *Eurytoma tylodermatis* Ashm. (Hymenoptera: Eurytomidae) from South Dakota. Proceedings of the Entomological Society of Washington 92(1): 96–100.

McDaniel, B. and A. Boe 1991b. A new *Bruchophagus* from *Glycyrrhiza lepidota* Pursh in the northern Great Plains (Hymenoptera: Chalcidoidea: Eurytomidae). Proceedings of the Entomological Society of Washington 93(3): 776–783.

McPherson, J.E. 1982. The Pentatomoidea (Hemiptera) of northeastern North America with emphasis on the fauna of Illinois. Southern Illinois University Press, Carbondale, IL. 240 p.

McPherson, J. E. and R. M. McPherson. 2000. Stink Bugs of Economic Importance in America North of Mexico. CRC LCC, Boca Raton, FL.

Metzler, E.H., J.A. Shuey, A. Ferge, R.A. Henderson and P.Z. Goldstein 2005. Contributions to the understanding of Tallgrass Prairie-dependent butterflies and moths (Lepidoptera) and their biogeography in the United States. Bulletin of the Ohio Biological Survey, New series, 15(1): viii + 143.

Miller, W.E. 1987. Guide to the Olethreutine Moths of Midland North America (Tortricidae). United States Department of Agriculture, Forest Service, Agriculture Handbook 660. 104 p.

Morgan, G.R. and R.R. Weedon. 1990. Oglala Sioux use of medicinal herbs. Great Plains Quarterly 10: 18–35.

Munyaneza, J. and J.E. McPherson. 1994. Comparative study of life histories, laboratory rearing, and immature stages of *Euschistus servus* and *Euschistus variolarius* (Hemiptera: Pentatomidae). The Great Lakes Entomologist 26(4): 263–274.

Muesebeck, C.F.W., K.V. Krombein and H.K. Townes. 1951. Hymenoptera of America North of Mexico Synoptic Catalog. United States Department of Agriculture Monograph No. 2. 1420 p.

NAMPG (North American Moth Photographers Group). 2023. *Hystrichophora ochreicostana* (Walsingham). http://mothphotographersgroup.msstate.edu/species.php?hodges=3396.

Noyes, J.S. 2019. Universal Chalcidoidea Database [last updated March 2019]. https://www.nhm.ac.uk/our-science/data/chalcidoids/database/index.dsml. (last accessed 24 January 2024).

Nuttall, T. 1818. The Genera of North American Plants, and a Catalogue of the species to the year 1817, volume II. Privately published, Philadelphia, PA. 254 + 10 [index] + [4, erratum (sic) & additions] p.

Ode, D.J. 2006. Dakota Flora: A seasonal sampler. South Dakota State Historical Society Press, Pierre, South Dakota. [ii] + 260 p.

Osborn, H. and H.A. Gossard. 1891a. The clover-seed caterpillar (*Grapholitha interstinctana*, Clem.) [pp. 166–169]. *In*: Reports on Entomological Work, Iowa Agricultural Experiment Station Bulletin No. 14. pp. 121–195.

Osborn, H. and H.A. Gossard. 1891b. The clover-seed caterpillar (*Grapholitha interstinctana*, Clem.) [pp. 260–262]. *In*: Reports on Injurious Insects, Iowa Agricultural Experiment Station Bulletin No. 15. pp. 197–296.

Owen, W.R. 1991. New host for *Acanthoscelides aureolus* Horn (Coleoptera: Bruchidae). The Pan-Pacific Entomologist 67(1): 73–74.

Panizzi, A. R. 1997. Wild hosts of pentatomids: Ecological significance and role in their pest status on crops. Annual Review of Entomology 42: 99–122.

Parish, H.E. 1934. Biology of *Euschistus variolarius* P. De B. (Family Pentatomidae; Order Hemiptera). Annals of the Entomological Society of America 27: 50–54.

Parshley, H.M. 1922. Report on a collection Hemiptera-Heteroptera from South Dakota. South Dakota State College, Technical Bulletin No. 2. 22 p.

Peck, O. 1963. A Catalogue of the Nearctic Chalcidoidea (Insecta: Hymenoptera). The Canadian Entomologist, Supplement 30. 1092 p.

Pezzini, D.T., C.D. DiFonzo, D.L. Fink, T.E. Hunt, J.J. Knodel, C.H. Krupke, B. McCornack, A.P. Michel, C.R. Philips, A.J. Varenhorst, R.J. Wright and R.L. Koch. 2019. Community composition, abundance, and phenology of stink bugs (Hemiptera: Pentatomidae) in soybean in the North Central Region of the United States. Journal of Economic Entomology 112(4): 1722–1731.

Pfaffenberger, G.S. and C.D. Johnson. 1976. Biosystematics of the first-stage larvae of some North American Bruchidae (Coleoptera). United States Department of Agriculture, Agricultural Research Service Technical Bulletin No. 1525. iv + 75 p.

Phillips, H.W. 2003. Plants of the Lewis & Clark Expedition. Mountain Press Publishing Company, Missoula, MT. vii + 277 p.

Pursh, F. 1814. Flora Americæ Septentrionalis; or, a systematic arrangement and description of the plants of North America. Containing, besides what have been described by preceding authors, many new and rare species, collected during twelve years travels and residence in that country. Volume II. White, Cochrane, and Co., London. pp. 359–751, 8 pl.

Ribeiro-Costa, C.S. and L.M. Almeida. 1981. Seed-chewing beetles (Coleoptera: Chrysomelidae, Bruchinae) [pp. 325–352]. *In*: V. Labeyrie (ed.), The Ecology of Bruchids Attacking Legumes (Pulses). Proceedings of the International Symposium held at Tours (France), April 16–19, 1980. Series Entomologica, vol. 19. Springer Dordrecht. xiv + 233 p.

Rice, M.E. 2015. Terry L. Erwin: She had a black eye and in her arm she held a skunk. American Entomologist 61(1): 9–15; reprinted: Rice, 2015, ZooKeys 500: 9–24.

Rider, D. A. 2012. The Heteroptera (Hemiptera) of North Dakota I: Pentatomomorpha: Pentatomoidea. The Great Lakes Entomologist 45(3–4): 312–380.

Rider, D.A. and D.R. Swanson. 2021. A distributional synopsis of the Pentatomidae (Heteroptera) north of Mexico, including new state and provincial records. Zootaxa 5015(1): 001–069.

Ritcher, P.O. 1966. Biological Control of Insects and Weeds in Oregon. Oregon State University Agricultural Experiment Station Technical Bulletin 90. 39 p.

RHNM (Red de Herbarios del Noroeste de México). 2023. *Glycyrrhiza lepidota* Pursh. https://herbanwmex.net/portal/taxa/index.php?taxon=1444&clid=3624.

Rogers, D.J. 1980. Lakota Names and Traditional Uses of Native Plants by Sicangu (Brule) People in the Rosebud Area, South Dakota: A Study Based on Father Eugene Buechel's Collection of Plants of Rosebud around 1920. Rosebud Educational Society, Inc., St. Francis, South Dakota. 111 p.

Schaefer, C.W. 1980. The host plants of the Alydinae, with a note on heterotypic feeding aggregations (Hemiptera: Coreoidea: Alydidae). Journal of the Kansas Entomological Society 53(1): 115–122.

Schaefer, C.W. and P.L. Mitchell 1983. Food plants of the Coreoidea (Hemiptera: Heteroptera). Annals of the Entomological Society of America 76: 591–615.

Schaefer, C.W. and A.R. Panizzi (eds.). 2000. Heteroptera of Economic Importance. CRC Press, Boca Raton, FL. [xii +] 828 p.

Schuh, R.T. and J.A. Slater. 1995. True Bugs of the World (Hemiptera: Heteroptera)Classification and Natural History. Cornell University Press, Ithaca, NY. xii + 336 + [1] p.

Scudder, G.G.E. 1997. True Bugs (Heteroptera) of the Yukon [pp. 241–336]. *In*: H.V. Danks and J.A. Downes (Eds.), Insects of the Yukon. Biological Survey of Canada (Terrestrial Arthropods), Ottawa. 1034 p.

SDDA (South Dakota Department of Agriculture). 2009. South Dakota Weeds (rev. ed.). South Dakota Department of Agriculture, Pierre. 235 p.

Stevens, O.A. 1963. Handbook of North Dakota Plants. North Dakota Institute for Regional Studies, Fargo. [2] + 237 p.

Swanson, D.R. 2018. New state records for some Pentatomomorpha (Heteroptera) of the United States. The Great Lakes Entomologist 51(3–4): 42–53.

Tropicos. 2023. *Glycyrrhiza astragalina* Gillies ex Hook. & Arn. https://www.tropicos.org/name/13019844.

Uhler, P.R. 1878. Notices of the Hemiptera Heteroptera in the collection of the late T.W. Harris, M.D. Proceedings of the Boston Society of Natural History 19: 365–446.

Underhill, G.W. 1943. Two pests of legumes: *Alydus eurinus* Say, and *A. pilosulus* Herrick-Schaeffer. Journal of Economic Entomology 36(2): 289–294.

Urbahns, T.D. 1914. The chalcis-fly in alfalfa seed. United States Department of Agriculture Farmers' Bulletin 636, 10 p.

Urbahns, T.D. 1919. Life-history observations on four recently described parasites of *Bruchophagus funebris*. Journal of Agricultural Research 16: 165–174, pl. 22–23.

Urbahns, T.D. 1920. The clover and alfalfa seed chalcis-fly. United States Department of Agriculture Bulletin No. 812, 20 p.

USDA Plants Database 2023. https://plants.usda.gov/home/plantProfile?symbol=GLLE3

Venugopal, P.D., P.L. Coffey, G.P. Dively, and W.O. Lamp. 2014. Adjacent habitat influence on stink bug (Hemiptera: Pentatomidae) densities and the associated damage at field corn and soybean edges. PloS One 9: e109917. https://doi.org/10.1371/journal.pone.0109917

Visher, S.S. 1914. The biology of Harding County northwestern South Dakota. Report of the State Geologist for 1913–1914. State Publishing Company, Pierre, SD. 126 p.

Wagner, D.L., E.M. Grames, M.L. Forister, M.R. Berenbaum and D. Stopak. 2021. Insect decline in the Anthropocene: Death by a thousand cuts. Proceedings of the National Academy of Sciences 118 No. 2: e2023989118. https://doi.org/10.1073/pnas.2023989118

Wahab, S., S. Annadurai, S.S. Abullais, G. Das, W. Ahmad, M.F. Ahmad, G. Kandasamy,R. Vasudevan, M.S. Ali, M. Amir, M. 2021. *Glycyrrhiza glabra* (Licorice): A Comprehensive Review on Its Phytochemistry, Biological Activities, Clinical Evidence and Toxicology. Plants 10, 2751: 36 pp. https://doi.org/10.3390/plants10122751

Walsingham, T. de G. 1879. Illustrations of typical specimens of Lepidoptera Heterocera in the collection of the British Museum. Part IV. – North-American Tortricidae. Taylor and Francis, London. xi + 84 p., pl. lxi–lxxvii.

Walsingham, T. de G. 1884. North American Tortricidae. The Transactions of the Entomological Society of London 1884: 121–147, pl. IV.

Weaver, J.E. 1954. North American Prairie. Johnsen Publishing Co., Lincoln, NE. xi + 348 p.

Weaver, J.E. 1968. Prairie Plants and their Environment: A Fifty-Year Study in the Midwest. University of Nebraska Press, Lincoln, Nebraska. xi + 276 p.

Weaver, J.E. and W.W. Hansen. 1941a. Native midwestern pastures, their origin, composition and degeneration. University of Nebraska Conservation and Survey Division Bulletin 22. 93 p.

Weaver, J.E. and W.W. Hansen 1941b. Regeneration of native midwestern pastures under protection. University of Nebraska Conservation and Survey Division Bulletin 23. 91 p.

Wehrle, L.P. 1924. The Clover-Seed Caterpillar. Cornell University Agricultural Experiment Station Bulletin 428, Ithaca, New York. 34 p.

Wen, J., Z.-L. Nie and S.M. Ickert-Bond. 2016. Intercontinental disjunctions between eastern Asia and western North America in vascular plants highlight the biogeographic importance of the Bering land bridge from late Cretaceous to Neogene. Journal of Systematics and Evolution 54(5): 469–490.

Wick, A.F., B.A. Geaumont, K.K. Sedivec and J.R. Hendrickson. 2016. Grassland Degradation [pp. 257–276]. *In*: J.F. Shroder and R. Sivankpillai (eds.), Biological and Environmental Hazards, Risks, and Disasters. Elsevier, Philadelphia. 492 p.

Willerslev, E., J. Davison, M. Moora, M. Zobl, E. Cloissac, M.E. Edwards, E.D. Lorenzen, M. Vestergård, G. Gussarova, J. Haile, J. Craine, L. Gielly, S. Boessenkool, L.S. Epp, P.B. Pearman, R. Cheddadi, D. Murray, K.A. Bråthen, N. Yoccoz, H. Binney, C. Cruaud, P. Wincker, T. Goslar, I.G. Alsos, E. Bellemain, A.K. Grysting, R. Elven, J.H. Sønstebø, J. Murton, A. Sher, M. Rasmussen, R. Rønn, T. Mourier, A. Cooper, J. Austin, P. Möller, D. Froese, G. Zazula, F. Pompanon, D. Rioux, V. Niderkorn, A. Tikhonov, G. Savvinov, R.G. Roberts, R.D.E. MacPhee, M.T.P. Gilbert, K.H. Kjær, L. Orlando, C. Brochmann and P. Taberlet. 2014. Fifty thousand years of Arctic vegetation and megafaunal diet. Nature 506: 47–51.

Wilkinson, J.D. and D.M. Daugherty. 1967. Biology of the broadheaded bug *Alydus pilosulus* (Hemiptera: Alydidae). Annals of the Entomological Society of America 60(5): 1018–1021.

Wimberly, M.C., L.L. Janssen, D.A. Hennessy, M. Luri, N.M. Chowdhury, and H. Feng. 2017. Cropland expansion and grassland loss in the eastern Dakotas: New insights from a farm-level survey. Land Use Policy 63: 160–173.

Wright, C.K. and M.C. Wimberly. 2013. Recent land use change in the western corn Belt threatens grasslands and wetlands. Proceedings of the National Academy of Science, U.S.A. 110: 4134–4139.

Wynia, R., A. Boe and W. Gardner. 1981. Initial investigation of chlorophyll deficient cotyledons of American licorice (*Glycyrrhiza lepidota* Pursh). Proceedings of the South Dakota Academy of Science 60: 173.

Xue, Z., T. Peng, B. Liu, Y. Liuy, Z. Zhang, K.A.G. Wyckhuys, P. Wang and Y. Lu. 2022. Licorice strips enhance predator-mediated biological control in Chinas's cotton crop. Pest Management Science 79(2): 781–791.

Yonke, T.R. and J.T. Medler 1968. Biologies of three species of *Alydus* in Wisconsin. Annals of the Entomological Society of America 61(2): 526–531.

A TINY DEER WITH BIG IMPLICATIONS: A NEW GENUS (*SANTUCCIMERYX*) FROM BADLANDS NATIONAL PARK HELPS BRIDGE THE GAP BETWEEN OLIGOCENE AND MIOCENE LEPTOMERYCIDAE (MAMMALIA, ARTIODACTYLA)

Mattison Shreero[1], Ed Welsh[1], Katherine Marriott[2,3], Donald R. Prothero[4,5]
[1] Badlands National Park, Interior, SD 57750
[2] Research Foundation of the City University of New York, New York, NY 10036
[3] American Museum of Natural History, New York, NY 10024
[4] Department of Geological Sciences, California State Polytechnic University, Pomona, CA 91768
[5] Department of Vertebrate Paleontology, Natural History Museum of Los Angeles County, Los Angeles, CA 90041

ABSTRACT

In 2016, a leptomerycid skull was collected at Badlands National Park, South Dakota from the earliest Whitneyan "*Leptauchenia* beds" (early Oligocene, ~32 mya) of the Poleslide Member, Brule Formation. This specimen (BADL 64015) is the first cranium referable to *Leptomeryx elissae* and is by far the most complete specimen to date; the skull lacks only the premaxilla and has associated dentaries. Prior to the discovery of BADL 64015, only isolated dentaries from the late Orellan of northwestern Nebraska had been referred to *L. elissae*.

Previous literature reports that *L. elissae* differs from the more common *Leptomeryx evansi* in that it has smooth tooth enamel, lacks the *Palaeomeryx*-fold on the lower molars, has a rounded and evident entoconulid on m3, and is smaller in size. Based upon measurements and diagnostic characters, BADL 64015 significantly deviates from *L. evansi* beyond differences in the lower teeth.

BADL 64015 exhibits a shift towards brachycephaly, a shortening of the cranium. The weakly developed postorbital constriction and reduced length of the sagittal crest demonstrate that the cranial characters of BADL 64015 are more akin to the Hemingfordian (early Miocene) leptomerycid, *Pseudoparablastomeryx*, than to *Leptomeryx*. This affinity suggests *L. elissae* cannot be attributed to the genus *Leptomeryx*. However, in BADL 64015, characters of the premolars are more closely aligned to *Leptomeryx* than *Pseudoparablastomeryx*, with some minor differences, including the shortening of P2-3. An enigmatic Arikareean (Oligocene to early Miocene) leptomerycid, *Pronodens*, also demonstrates a shortening of the anterior dentary, but lacks associated cranial material. The *Pronodens* holotype consists of only extremely worn lower teeth, making it impossible to draw any meaningful comparisons between *Pronodens* and other leptomerycids.

Based on the combination of cranial and dental characteristics, we thus validate *L. elissae* as a distinct species and reposition it into a new genus, *Santuccimeryx*, intermediate between *Leptomeryx* and *Pseudoparablastomeryx*. Evolutionary relationships of the Leptomerycidae will, however, remain uncertain until additional specimens can be collected that provide unambiguous characters to clarify the relationships between these taxa.

Zoobank.org registration LSID: urn:lsid:zoobank.org:pub:212B0DF4-20E4-4A2C-A7BF-F1224274921E **Published on** 29 Month 2024

Keywords

Leptomerycidae, *Leptomeryx, Pseudoparablastomeryx, Santuccimeryx,* Brule Formation, Badlands National Park, South Dakota

INTRODUCTION

A new specimen (BADL 64015), initially attributed to *Leptomeryx*, was discovered by a Geoscientists-in-the-Parks intern at Badlands National Park during the summer of 2016 (Figure 1). After preparation, we used dental characters to identify the specimen as *Leptomeryx elissae*, a species named by Korth and Diamond (2002) based on isolated dentaries and lower dentition (Figure 2). BADL 64015 is the first and only known specimen of *L. elissae* to contain both dentary and skull material, thereby providing the first characters beyond lower dentition and helping clarify evolutionary relationships within the genus *Leptomeryx* and the family Leptomerycidae. This work describes BADL 64015 and compares it to the only other leptomerycid genera known from skull material: *Leptomeryx* and *Pseudoparablastomeryx*.

The ruminant family Leptomerycidae is an extinct taxon endemic to North America and the longest-lived member of the Traguloidea. Leptomerycids range temporally from the late Eocene to Miocene and geographically throughout the Western Interior from Saskatchewan to Mexico, with one atypical occurrence in California (*Leptomeryx blacki*). Leptom-

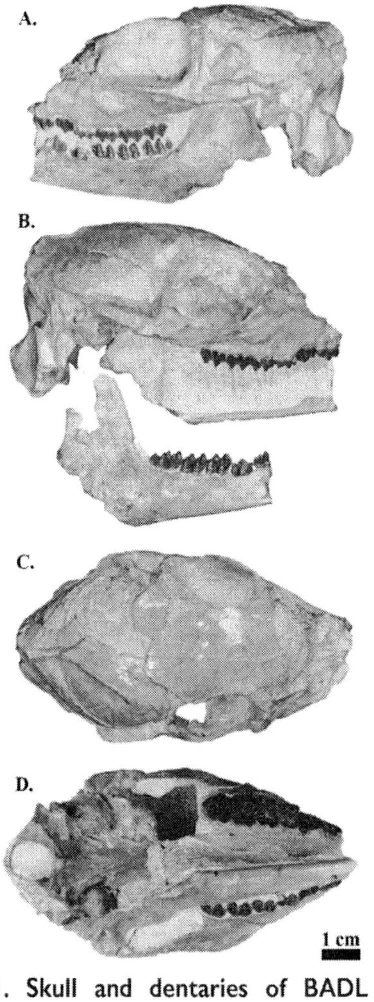

Figure 1. Skull and dentaries of BADL 64015, the first skull attributed to the species *Leptomeryx elissae* (Korth and Diamon 2002). A. left lateral view; B. right lateral view; C. dorsal view; D. ventral view.

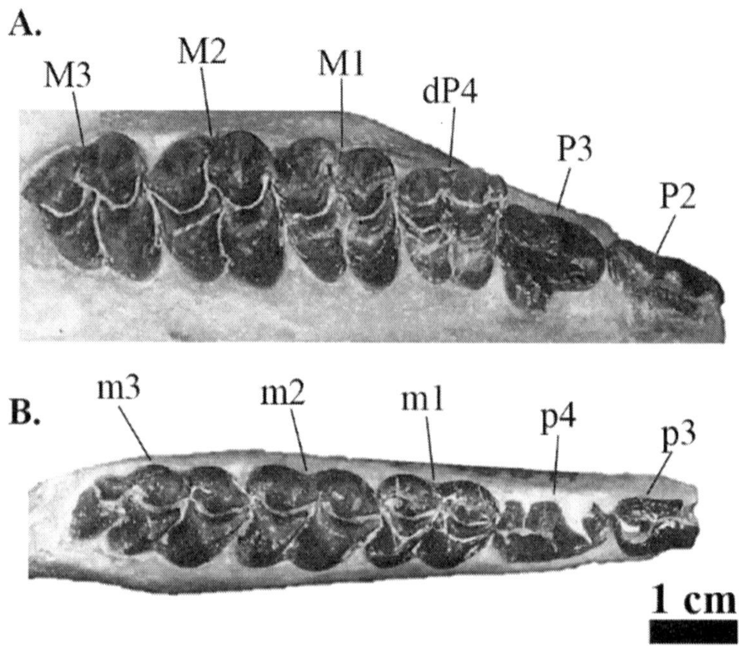

Figure 2. Dentition of BADL 64015. A. occlusal view of upper dentition; B. occlusal view of lower dentition. Lower dentition was used to initially identify BADL 64015 as *Leptomeryx elissae* (Korth and Diamond 2002).

erycidae currently contains four genera of small (~2 to 4 kg), hornless ruminants: *Hendryomeryx, Leptomeryx, Pronodens,* and *Pseudoparablastomeryx.* Of these, *Leptomeryx* is by far the most abundant genus and is also one of the most common small herbivores found in the White River Group.

Members of this genus are extremely useful in biostratigraphy. All three Eocene *Leptomeryx* species have their First Appearance Datum (FAD) in successive subdivisions, or interval zones, of the Chadronian North American Land Mammal Age (NALMA). The late early Chadronian *Leptomeryx yoderi* Interval Zone (Ch2) is followed by the *Leptomeryx mammifer* Interval Zone (Ch3) and the *Miniochoerus chadronensis* Interval Zone (Ch4) for which the last Chadronian leptomerycid, *Leptomeryx speciosus*, can be used as an index fossil (Prothero and Whittlesey 1998, Prothero and Emry 2004). The transition into the Orellan NALMA (Or1; *Hypertragulus calcaratus* Interval Zone) at the Eocene–Oligocene transition is marked by the replacement of *L. speciosus* with *Leptomeryx evansi* (Prothero and Emry 2004; see also Zanazzi et al. 2009). *L. evansi* differs from its Chadronian counterparts in that it possesses crenulated enamel and *Palaeomeryx*-folds on the lower molars (Figure 3A) (Heaton and Emry 1996, Zanazzi et al. 2009). The evolution of the *Palaeomeryx*-fold and relative proliferation of *Leptomeryx evansi* throughout the White River Group of the northern Great Plains makes it an invaluable taxon in orienting the boundary of the Eocene–Oligocene transition.

Despite their abundance and relevance to biostratigraphy, the taxonomic organization of leptomerycids has confounded researchers for more than a century, and their evolutionary relationships remain primarily unresolved in the literature. Zittel (1893) was the first to distinguish Leptomerycinae as a distinct subfamily

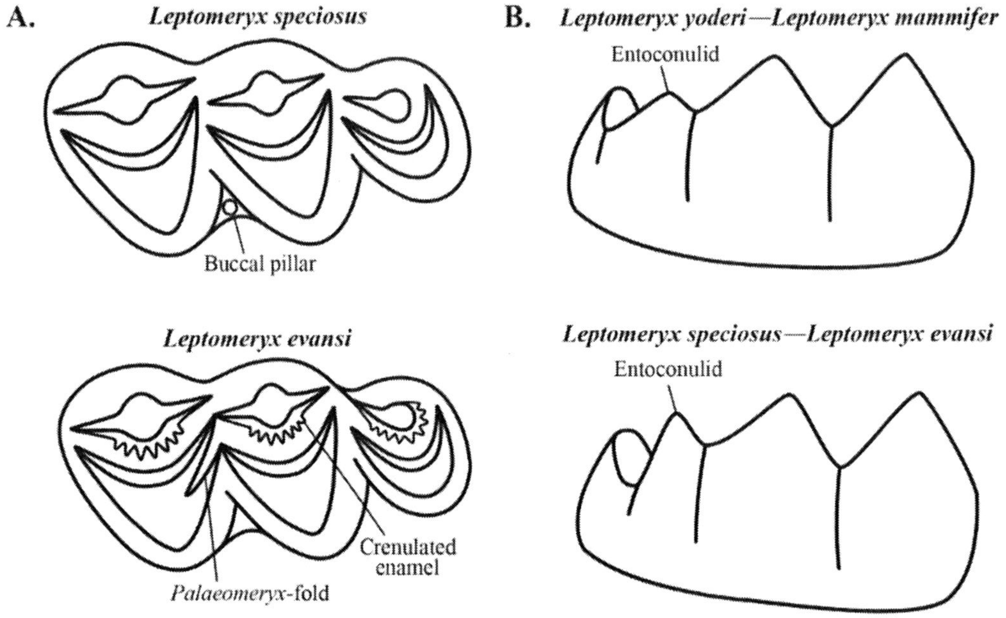

Figure 3. A. Idealized comparative diagram highlighting diagnostic molar features of *L. speciosus* (Chadronian) and *L. evansi* (Orellan). The *Palaeomeryx*-fold and crenulated enamel of *L. evansi* are used to indicate the beginning of the Orellan NALMA. B. Idealized comparative diagram of m3 entoconulid morphology illustrating the two recognized lineages of Leptomeryx: the *L. yoderi–L. mammifer* lineage and *L. speciosus–L. evansi* lineage. Modified from Heaton and Emry (1996) and Korth and Diamond (2002).

within the Tragulidae. Shortly thereafter, Scott (1899) first utilized Leptomerycidae as a distinct family, but provided no justification and was dismissed by subsequent authors who placed members within the Hypertragulidae (Frick 1937, Scott 1940, Simpson 1945, Macdonald 1963, 1970). Scott (1899) also included hypertragulids (*Hypertragulus* and *Hypisodus*) and protoceratids (*Protoceras*) within the family Leptomerycidae. Character differences to distinguish Leptomerycidae as a distinct family were not described until Gazin (1955), who focused on dental characters. However, Gazin's "leptomerycids" are no longer considered members of the family and are instead early members of the Protoceratidae and Camelidae (Wortman 1898, Matthew 1903, Scott 1899, Taylor and Webb 1976). Taylor and Webb (1976) provided the first cranial and postcranial characteristics, alongside dental characters, to further justify the distinction of the family Leptomerycidae, and this justification was supported by subsequent works (Webb and Taylor 1980, Webb 1998, Vislobokova 2001). Taylor and Webb (1976) also document the survival of two leptomerycid genera—*Pronodens* and *Pseudoparablastomeryx*—into the late Oligocene and Miocene, respectively. To clarify relationships within Hypertraguloidea and Traguloidea, Vislobokova (2001) elaborated on diagnostic characters through phylogenetic analyses and repositioned taxa previously and erroneously placed within Leptomerycidae (e.g., *Archaeomeryx*). Currently, taxa that can be confidently included in Leptomerycidae are *Hendryomeryx*, *Leptomeryx*, *Pronodens*, and *Pseudoparablastomeryx*. Another genus, *Pipestoneia*, was tentatively referred to Leptomerycidae (Tabrum and Métais 2007); however, no further study had been undertaken to resolve the position of this taxon relative to other leptomerycids or tragulids. Métais and Vislobokova (2007) further expanded upon phylogenetic relationships of basal ruminants, reaffirming leptomerycids within the Traguloidea. Spaulding et al. (2009) provided a phylogenetic analysis of Artiodactyla, focusing on Cetacea, and added confusion by placing *Leptomeryx* within the Pecora as the sister group to Moschidae. However, this work did not include basal, hornless ruminants accounted for in other studies which consistently place *Leptomeryx* close to the Tragulidae (Webb and Taylor 1980, Bouvrain and Geraads 1985, Geraads et al. 1987, Janis 1987, Gentry and Hooker 1988, Vislobokova 2001, Métais and Vislobokova 2007).

Heaton and Emry (1996) provided the first thorough analyses of the genus *Leptomeryx*, employing a large sample size of Chadronian specimens from Wyoming and Montana. They statistically accounted for every character variation that could be found within the genus, provided a synopsis of reliable characterization for the four species they considered valid (*L. yoderi*, *L. mammifer*, *L. speciosus*, and *L. evansi*), and analyzed which characters could be attributed to individual variation. As with most other works on the genus, Heaton and Emry (1996) relied on characters from the lower dentition because upper dentitions are considered less diagnostic and are rarely found with associated dentaries. Heaton and Emry's (1996) diagnostic characters are based on size, morphology of the m3 entoconulid, and presence or absence of the *Palaeomeryx*-fold. Korth and Diamond (2002) continued the reliance on dental characters in their reevaluation of Orellan *Leptomeryx* species when they resurrected *L. exilis* (Cook, 1934) and named a new species, *Leptomeryx elissae*. Mathis and MacFadden (2010) used enamel thickness and complexity data to verify species separation along the Eocene–Oligocene transition, corroborating Korth and Diamond's (2002) claim that there was at least one other species of leptomerycid in addition to *L. evansi* during the Oligocene.

Heaton and Emry (1996) proposed two evolutionary lineages, simplified as the *L. yoderi–L. mammifer* and *L. speciosus–L. evansi* lineages, based on the morphology of the m3 entoconulid. The m3 entoconulid either possesses a posterior ridge (*L. yoderi–L. mammifer*) or the ridge is lost and the m3 entoconulid is a pillar-like cuspid (*L. speciosus–L. evansi*) (Figure 3B). Korth and Diamond (2002) elaborated on these lineages with their analysis of *L. elissae* and *L. exilis*. They defined *L. elissae* based on a left dentary containing p3-m3 (UNSM 119638) with a unique combination of characters–smooth tooth enamel, *L. speciosus–L. evansi* type of m3 entoconulid, absence of a *Palaeomeryx*-fold on the lower molars, and smaller size than *L. evansi*–and provided several referred specimens. Thus, *L. elissae* seems to have stemmed from the *L. speciosus–L. evansi* lineage since it shares the same m3 morphology, whereas *L. exilis* shares the m3 morphology of the *L. yoderi–L. mammifer* group and implies a ghost lineage somewhere in the early Orellan NALMA (Korth and Diamond 2002). Korth and Diamond (2002) also elaborated on the relationship of *L. obliquidens* (Lull 1922), a species not discussed by Heaton and Emry (1996). They position *L. obliquidens* within the *L. yoderi–L. mammifer–L. exilis* lineage on the basis of the m3 entoconulid and lack of *Palaeomeryx*-fold but note that *L. obliquidens* has highly crenulated enamel, a trait typical of the *L. speciosus–L. evansi* lineage.

Considering the confusion surrounding the evolutionary relationships of the Leptomerycidae, this work endeavors to not only describe new material of *L. elissae*, but also compare it to all leptomerycid genera known from skull material in order to provide an impartial analysis of the position of *L. elissae* within Leptomerycidae. As mentioned, only the genera *Hendryomeryx*, *Leptomeryx*, *Pronodens*, and *Pseudoparablastomeryx* are considered valid leptomerycids (Webb 1998, Vislobokova 2001). *Hendryomeryx* contains three species: *H. defordi, H. wilsoni,* and *H. esulcatus.* The holotype of *Hidrosotherium transpecosensis* (Wilson 1974) was synonymized with *H. defordi* (Prothero and Emry 1996) and is the only reported skull of any *Hendryomeryx* species. The characters in *Hendryomeryx* are basal and differ enough from the other leptomerycids that they do not merit detailed comparison to *L. elissae* for the purposes of this study. Aside from *L. elissae*, *Leptomeryx* contains the species *L. blacki*, *L. yoderi*, *L. mammifer*, *L. speciosus*, *L. evansi*, *L. exilis,* and *L. obliquidens.* Of these, only *L. evansi*, *L. obliquidens,* and now *L. elissae* are known from skull material. However, the poor condition of the only known skull of *L. obliquidens* precludes it from reliable comparison. Two other species, *L. transmontanus* and *L. agatensis*, have been plagued with vague reclassifications and require additional review. *Pronodens* contains one species, *P. silberlingi*, which is known from an isolated dentary with very worn teeth. Though no *Pronodens* skull material has been reported, this problematic genus merits discussion due to its temporal intermediacy between *Leptomeryx* and *Pseudoparablastomeryx* and its lack of species-defining characters. All referrals of material to this genus are tentative at best. *Pseudoparablastomeryx* contains the species *P. scotti* and *P. francescita.* Of these, *P. scotti* is known from a single skull (Frick 1937, Taylor and Webb 1976). As such, *Leptomeryx* and *Pseudoparablastomeryx* are currently the only leptomerycid genera with defined cranial characters, meriting their detailed comparison to BADL 64015.

Institutional Abbreviations—AMNH, American Museum of Natural History, New York, New York; BADL, Badlands National Park, Interior, South Dakota; CM, Carnegie Museum of Natural History, Pittsburgh, Pennsylvania; F:AM, Frick Collection, American Museum of Natural History, New York, New York; SDSM, South Dakota School of Mines and Technology Museum of Geology, Rapid City, South Dakota; UNSM, University of Nebraska State Museum, Lincoln, Nebraska; YPM, Yale Peabody Museum of Natural History, New Haven, Connecticut; YPM-PU, Princeton University Museum of Natural History (now in the collection of the Yale Peabody Museum of Natural History).

GEOLOGICAL SETTING AND PALEOENVIRONMENT

Badlands National Park is characterized by the White River Group siltstones and mudstones, which were deposited in the northern Great Plains between 37 and 30 million years ago (Figure 4). The White River Group contains two formations, the late Eocene Chadron Formation and the early Oligocene Brule

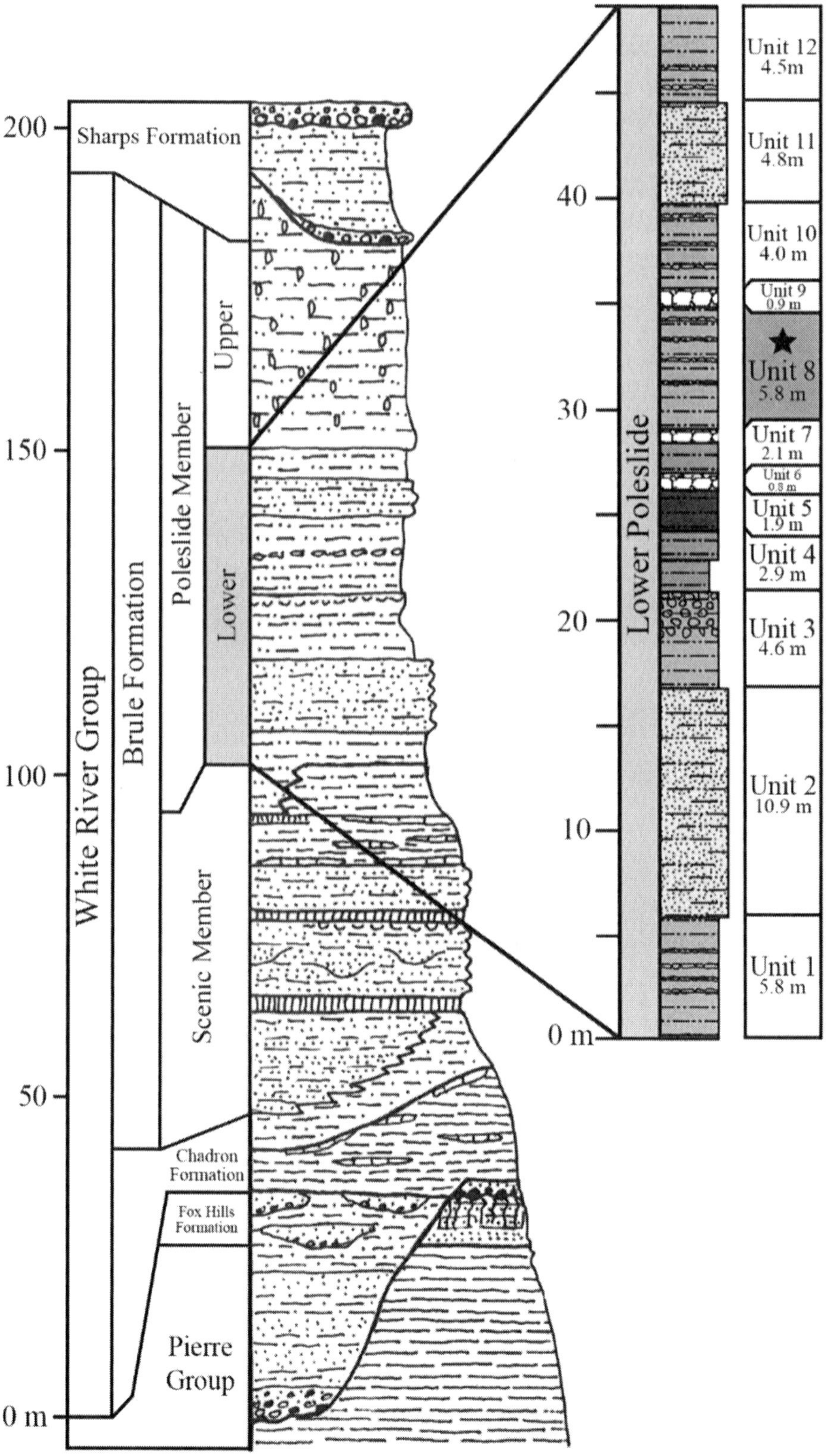

Figure 4. General stratigraphic column of Badlands National Park and inset showing detailed stratigraphy of the Lower Poleslide Member. BADL 64015 was found within Unit 8 of the Poleslide. Modified from Evanoff et al. (2010).

Formation. The Brule Formation in South Dakota is subdivided into the Scenic Member (34–32 mya) and Poleslide Member (32–30 mya). In the Cedar Pass area of the Badlands, the lower Poleslide is primarily lithologically composed of tan to light gray eolian siltstone but also contains mudstone and well-bedded river channel sandstone complexes (Benton et al. 2015). The specimen of *L. elissae* described herein (BADL 64015) was found in the Cedar Pass area within Unit 8 of the Poleslide Member, which has also been referred to as the Bjork Siltstone or the *Leptauchenia* Beds (Figure 4) (Evanoff et al. 2010). Biostratigraphically, Unit 8 correlates with the base of the *Leptauchenia major* Interval Zone as defined by Prothero and Whittlesey (1998) and Prothero and Emry (2004) because it contains the zone's characteristic fauna. Stevens et al. (2023) provides a synopsis of oreodont zonation that narrows the Orellan–Whitneyan transition in the Cedar Pass Area, which corroborates rodent data supplied by Korth (2014). Unit 8 thus records the early Oligocene, about 32 mya, and the earliest Whitneyan NALMA (Wh1).

In stark contrast to wetter, more tropical Eocene environments, around 32 mya the area that is now Badlands National Park was an open savanna with few trees. Paleosols of the lower Poleslide indicate that the environment here was a dry, distal floodplain (Benton et al. 2015). The prolific Cedar Pass fossil assemblage reveals a diverse collection of Cenozoic animals that roamed this landscape including, most commonly, oreodonts (*Leptauchenia* and *Merycoidodon*), hyracodontids (*Hyracodon*), ancient dogs (*Archaeocyon* and *Mesocyon*), false saber-toothed cats (*Nimravus*), tortoises (*Stylemys*), and small deer-like artiodactyls (*Leptomeryx* and *Hypertragulus*).

SYSTEMATIC PALEONTOLOGY

Class MAMMALIA Linnaeus 1758
Order ARTIODACTYLA Owen 1848
Family LEPTOMERYCIDAE Scott 1899
Genus *SANTUCCIMERYX* new genus
Santuccimeryx elissae (Korth and Diamond 2002)

Leptomeryx elissae Korth and Diamond 2002

Holotype—UNSM 119638, mandible with left p3-m3 (Korth and Diamond 2002).

Etymology—"Santucci's ruminant". Named in honor of Vincent L. Santucci, the Senior Paleontologist at the Geologic Resources Division of the National Park Service, for his contributions to NPS paleontology, support of and advocacy for the paleontology program at Badlands National Park, and work conducted in the Cedar Pass area for his master's thesis.

Referred Specimens—Specimens referred to *L. elissae* by Korth and Diamond (2002) from the Orella Member, Brule Formation, Sioux County, Nebraska: CM 73416; UNSM 119636, 119637, 119639, 119640; AMNH 53626A, 606-25868, 606-25873.

From the Poleslide Member, Brule Formation, Jackson County, South Dakota: BADL 64015, skull and dentaries with left and right P2-3, dP4, M1-3, left and right p3-4, m1-3, and articulated atlas; BADL 24824, partial skull with left P3-M3 and right P4-M2; BADL 64245, partial skull and dentaries with right dP4, M1-3, left and right dp4, m1-3.

Locality and Horizon—Holotype from UNSM locality Sx-8, Orella Member (Orella C; Or3) of the Brule Formation, Sioux County, Nebraska. Other specimens referred to by Korth and Diamond (2002) are from the upper Orella Member (Orella C and D; Or3 and Or4), Brule Formation, Sioux County, Nebraska.

New BADL material is from the Cedar Pass area, North Unit, Badlands National Park, Jackson County, South Dakota. These specimens were collected from Unit 8 of the Poleslide Member, Brule Formation, White River Group, based on Evanoff et al. (2010) (Figure 4). This unit is classically known as the base of the "*Leptauchenia* beds" (Schultz and Falkenbach 1968). Specific site information is protected under federal law. To request precise locality data, qualified researchers should inquire with Badlands National Park's on-staff paleontologist.

Distribution—Late Orellan of Sioux County, Nebraska and earliest Whitneyan of Jackson County, South Dakota.

***Emended Diagnosis*—Cranium:** *Santuccimeryx* is brachycephalic, exhibiting a short cranium and relatively large orbits. In BADL 64015, the orbit length is 61.2% of the cranium length measured from the postorbital process of the frontal. In *Santuccimeryx*, the postorbital bar is formed primarily by the jugal. The postorbital constriction is shallow and poorly developed. The distance between frontal-parietal contact and sagittal crest is double the length of the sagittal crest. The sagittal crest has weak insertions for nuchal ligaments. The angle between the temporal crests is 64°, and they converge with the anterior sagittal crest dorsal to the external auditory meatus. The sagittal crest does not overhang the occipital region. The basioccipital is posteriorly expanded and trapezoidal, and the auditory bullae are sculptured, defined, and possess an anterior spur.

Dentition*:* *S. elissae* has smooth tooth enamel. P2 and P3 have distinct preprotocrista and lack a cusp-like protocone. The margin of the P3 postprotocrista is posterolabially expanded, and the P3 posterostyle is reduced. BADL 64015 possesses upper deciduous P4s, but lower deciduous teeth are lost, and all upper and lower molars are fully erupted, indicating an adult specimen. The dP4 is strongly molariform, nearly as robust as M1, and possesses a lingual cingulum and buccal pillar. The styles of all upper molars are reduced or nearly absent and antero-posteriorly shortened.

The lower premolars are complex. The p3 possesses a protoconid with posterior ridges to the entoconid and hypoconid, and p4 has a well-developed paraconid, metaconid, protoconid, entoconid, and hypoconid. The dp4 in referred *S. elissae* specimen BADL 64245 is molariform with three characteristic lobes, as in all artiodactyls. On the lower molars, *S. elissae* has smooth enamel, anterior cingula on the protoconids, and lacks *Palaeomeryx*-folds. The posterior ridge on m3 is lost and the entoconulid is a pillar-like cuspid (*L. speciosus–L. evansi* morphology).

Differential Diagnosis—Overall, the cranial features of BADL 64015 are closest to *Pseudoparablastomeryx*, while the dental characters are more akin to

TABLE 1 – Differential characters among all leptomerycids observed in this study. "✓" indicates character is present, "X" indicates character is absent, "?" indicates character is unknown.

	Palaeomeryx-Fold	Crenulated Enamel	m3 Entoconulid Ridge	Buccal Pillar	Elongate P2-3 Posterior Ridge	Posterior Basin in P2-3 Postprotocrista	Well-Developed Upper Molar Styles
Hendryomeryx	X	X	?	✓	✓	X	?
Leptomeryx yoderi	X	X	✓	X	?	?	?
Leptomeryx mammifer	X	X	✓	X	?	?	?
Leptomeryx exilis	X	X	✓	X	✓	X	✓
Leptomeryx obliquidens	X	X	✓	X	✓	X	✓
Leptomeryx speciosus	X	X	X	✓	?	?	X
Leptomeryx evansi	✓	✓	X	X	✓	X	✓
Santuccimeryx elissae	X	X	X	✓	X	✓	X
Pseudoparablastomeryx	X	X	X	✓	X	✓	X
"Leptomeryx" transmontanus	?	X	?	?	X	?	?
"Leptomeryx" agatensis	?	X	X	X	✓	X	✓
Pronodens silberlingi	?	X	?	X	?	?	?
Pipestonia	X	✓	X	X	✓	X	✓

	Postorbital Bar Formed Primarily By Jugal	Well-Developed Postorbital Constriction	Ratio of Sagittal Crest Length to Distance Between Frontal-Parietal Suture is ≤ 1/2	Temporal Crests Converge Above External Auditory Meatus	Angle of Temporal Crests ≥ 60°	Basioccipital is trapezoidal	Spherical Auditory Bulla
Leptomeryx evansi	X	✓	X	X	X	X	✓
Santuccimeryx elissae	✓	X	✓	✓	✓	✓	X
Pseudoparablastomeryx	X	X	✓	✓	✓	?	X

Figure 5. Artistic reconstruction of *Leptomeryx* (left) meeting *Santuccimeryx* (right). Note the overall smaller size, proportionally larger orbits, and brachycephaly of *Santuccimeryx*. Illustration by Mattison Shreero.

TABLE 2 – Cranial measurements of *Santuccimeryx elissae*, *Leptomeryx evansi*, and *Pseudoparablastomeryx scotti*. All measurements in mm unless otherwise notated.

	BADL 64015 *Santuccimeryx elissae*	SDSM 62381 *Leptomeryx evansi*	BADL 53936 *Leptomeryx* cf. *evansi*	F:AM 95490* *Pseudoparablastomeryx scotti*
Maximum Width of Cranium	33.5	33.8	28.4	35.9
Cranial Length From Postorbital Process of Frontal	36.4	44.7	45.8	38.3
Width at Postorbital Constriction	29.9	23.1	19.3	27.9
Length of Sagittal Crest	10.3	30.5	30.8	11.3
Distance Between Frontal-Parietal Suture & Sagittal Crest	22.0	7.4	9.5	18.9
Angle Between Temporal Crests	64°	44°	40°	60°
Depth at Malar	12.0	14.0	18.0	N/A
Height of Orbit	17.4	19.2	17.0	N/A
Length of Orbit	22.3	N/A	18.2	N/A

* F:AM 95490 exhibits some dorsal–ventral crushing

Leptomeryx, specifically *L. speciosus* (Table 1). Therefore, this taxon cannot be confidently placed within either genus, justifying repositioning *L. elissae* into a new, intermediate genus: *Santuccimeryx*. See figure 5 for artistic reconstruction of both *Leptomeryx* and *Santuccimeryx*.

***Comparison to Leptomeryx*:** The orbit in *Santuccimeryx* is proportionately larger than observed in *Leptomeryx* (Table 2). The orbit length in BADL 53936 (*L.* cf. *evansi*) is 39.7% of the cranium length measured from the postorbital process of the frontal whereas this ratio is 61.2% in BADL 64015 (*S. elissae*). In *L. evansi* (and in all leptomerycids known from cranial material), the postorbital bar is composed of equal proportions of the frontal and jugal. In *Santuccimeryx*, the jugal comprises the majority of the postorbital bar. The postorbital constriction is shallow and poorly developed, but well-developed in *L. evansi* (Figures 6 and 7). The cranium is shortened in *Santuccimeryx* with a reduced sagittal crest length of 10.3 mm; *L. evansi* exhibits a 33.8 mm long sagittal crest in SDSM 62381, a specimen of nearly identical cranial width (Table 2 and Figure 7). The shortening of the sagittal crest increases the distance between the frontal-parietal contact and the anterior margin of the sagittal in *Santuccimeryx* (Figures 6 and 7; Table 2). The ratio of these measurements is inversely proportional in observed specimens of *L. evansi* (SDSM 62381 and BADL 53936) (Table 2). The angle between temporal crests is 64° in contrast to 40-44° in *L. evansi* (Table 2). In *Santuccimeryx*, the anterior sagittal crest begins to form dorsal to the external auditory meatus; in *L. evansi*, the anterior sagittal crest is formed anterior to the lateral expansion of the squamosal for the zygomatic (Figures 6 and 7). The basioccipital in *Santuccimeryx* is trapezoidal whereas it is subrectangular in *L. evansi*.

In previous literature, differentiating species of *Leptomeryx* relies entirely on lower dentition (Heaton and Emry 1996, Korth and Diamond 2002). BADL 64015, BADL 24824, and BADL 64245 provide the first upper dentition associated with the species *S. elissae* (Figure 2). P3 is shortened in *S. elissae*, with the protocone in a medial position (Figure 8). In *L. evansi*, the posterior portion of

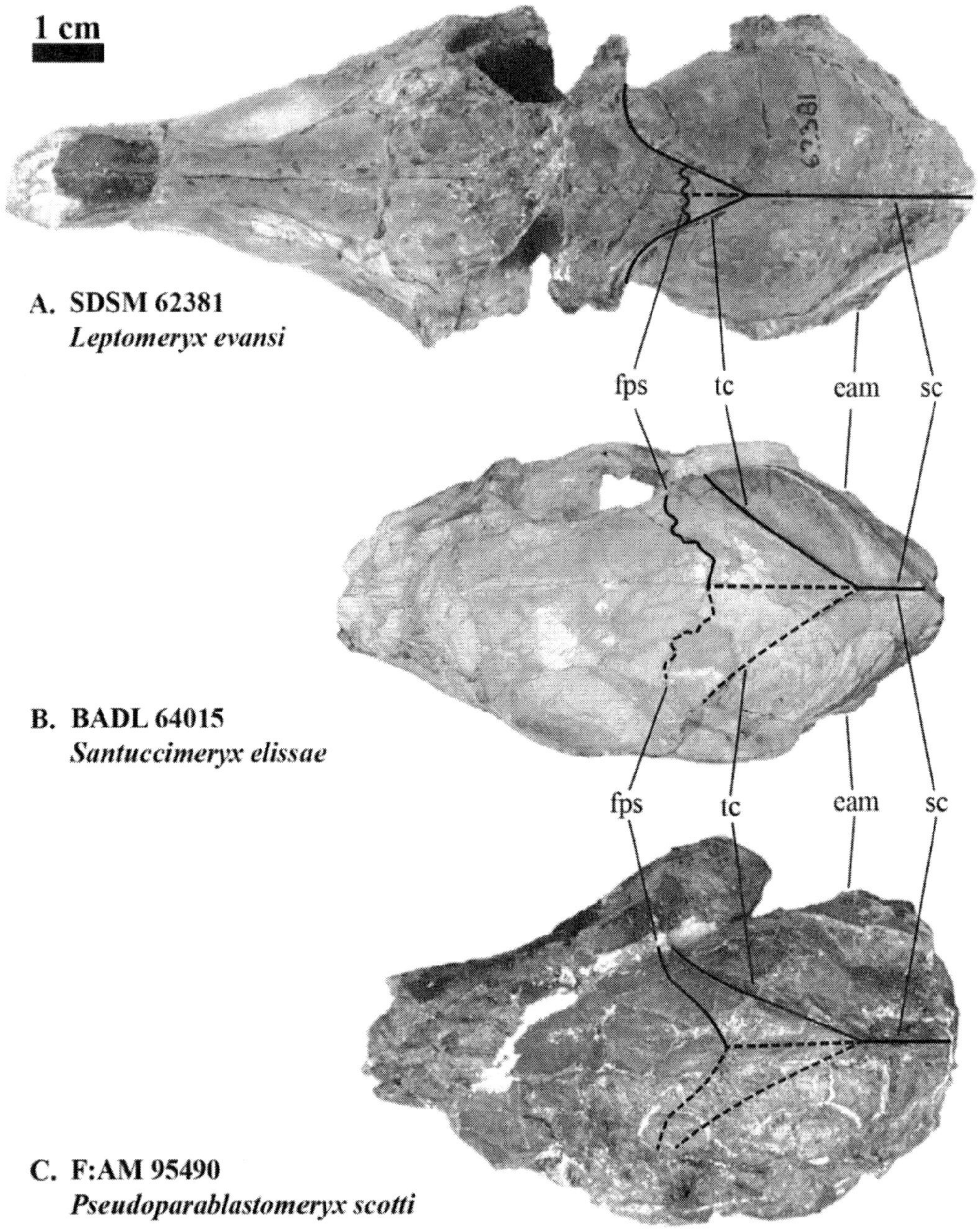

Figure 6. Comparative dorsal views of the crania of *Leptomeryx*, *Santuccimeryx*, and *Pseudoparablastomeryx*. Lines mark the positions and orientation of the frontal-parietal suture (fps), temporal crests (tc), external auditory meatus (eam), and sagittal crest (sc). *Santuccimeryx elissae* has a cranial anatomy more akin to *Pseudoparablastomeryx* than *Leptomeryx*. In BADL 64015 and F:AM 95490, note the shortened crania, shortened sagittal crests, increased distance between the frontal-parietal sutures and sagittal crests, and increased angle between temporal crests.

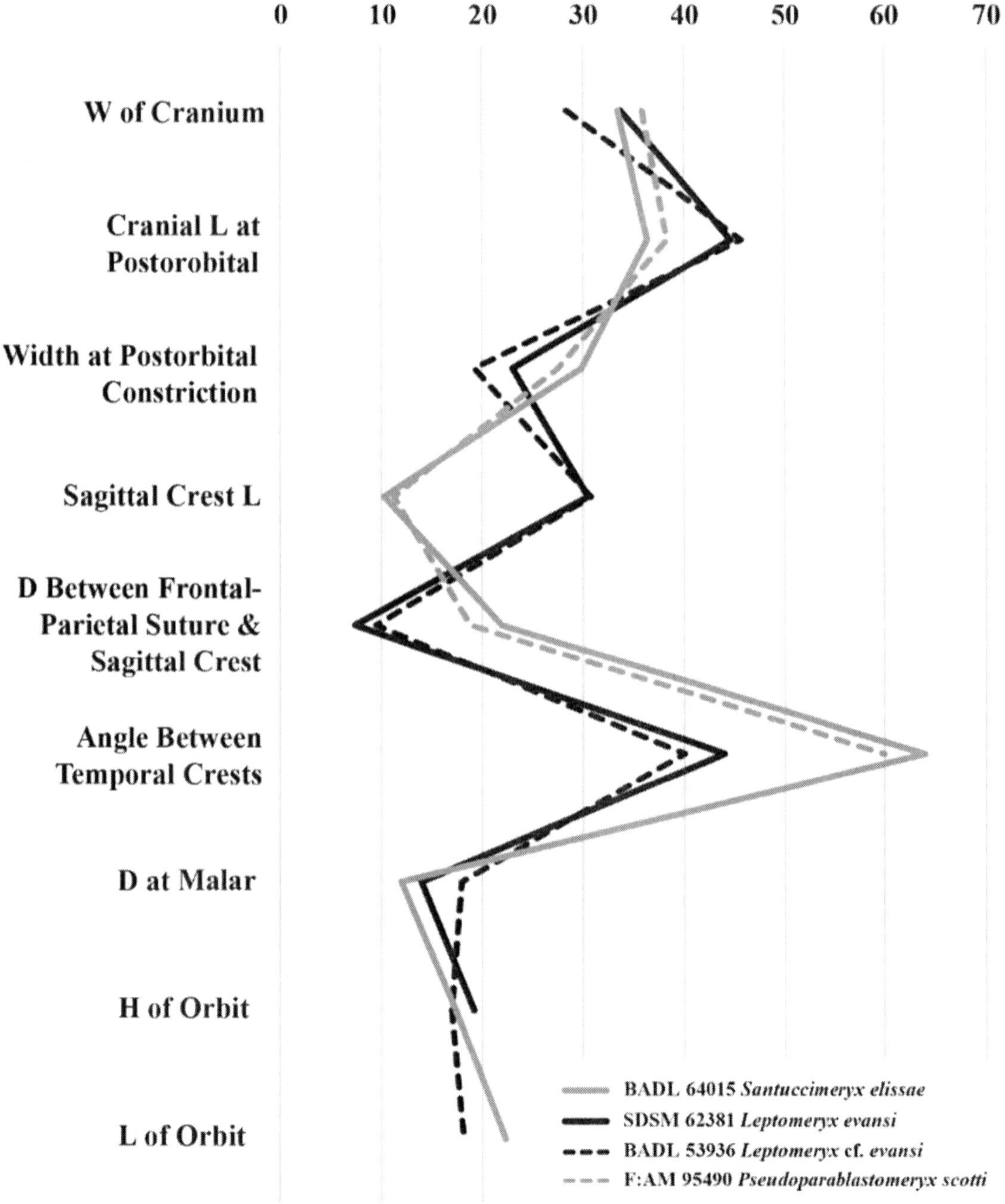

Figure 7. Simpson-ratio diagram visualizing the variation of cranial measurements for specimens of *Santuccimeryx*, *Leptomeryx*, and *Pseudoparablastomeryx*. Angle between temporal crests measured in degrees; all other measurements in mm. Note the significant deviation of cranial length at postorbital, sagittal crest length, and angle between temporal crests in BADL 64015 and F:AM 95490.

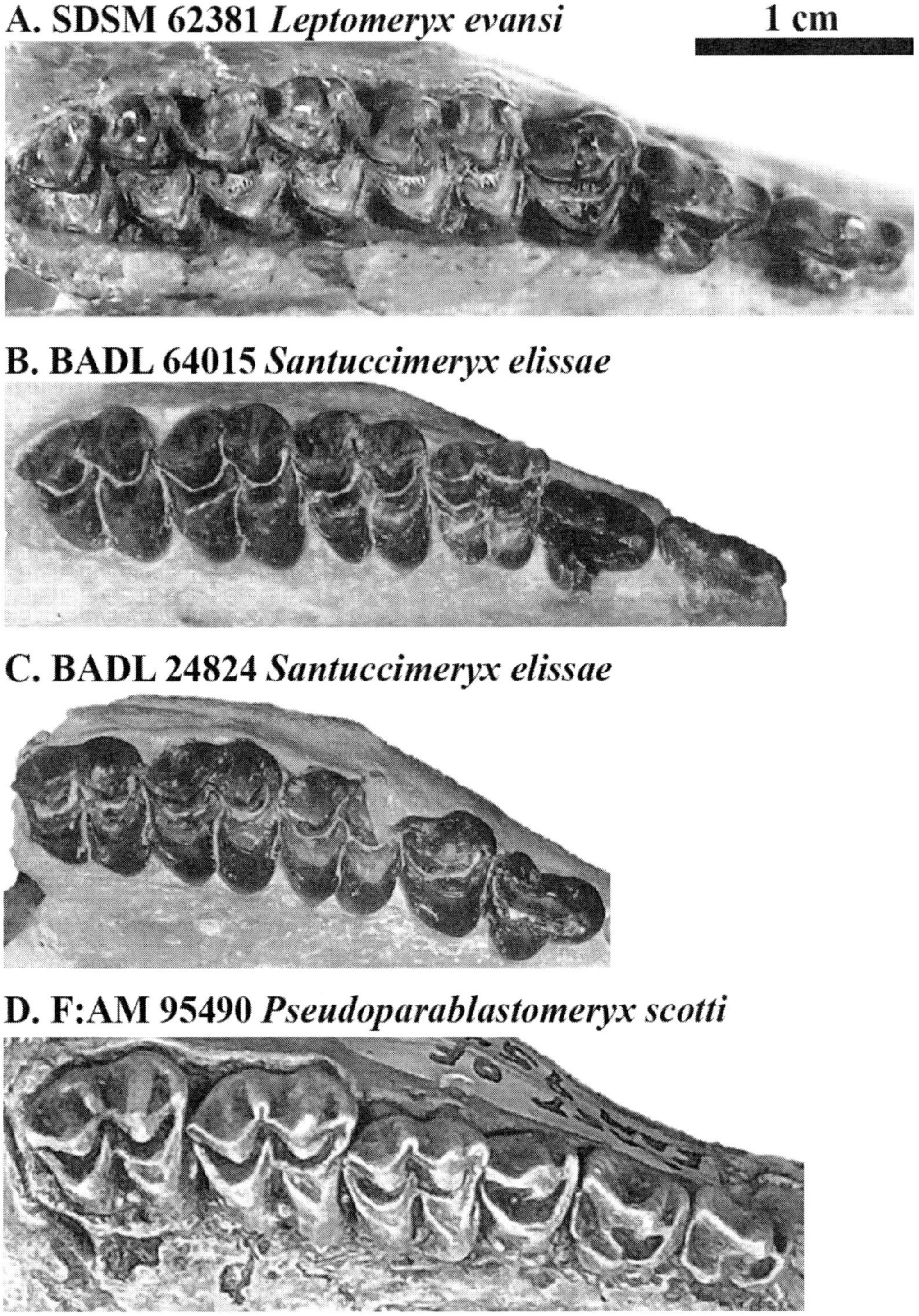

Figure 8. Comparative occlusal views of upper dentition for *Leptomeryx*, *Santuccimeryx*, and *Pseudoparablastomeryx* (cast of F:AM 95490). SDSM 62381, BADL 64015, and F:AM 95490 are right maxilla. BADL 24824 is a left maxilla (image has been flipped horizontally for ease of comparison). Note the submolariform P2-3 in F:AM 95490 in contrast to *Santuccimeryx* and *Leptomeryx* and the significantly increased molar styles of SDSM 62381 when compared to *Santuccimeryx* and *Pseudoparablastomeryx*.

P3 is more elongate and the protocone is in a more anterior position (Figure 8). The P3 postprotocrista is directed posterolabially, creating a small basin, whereas in *L. evansi* it conforms along the margins of the tooth in a sinuous pattern, behaving more like a cingulum (Figure 8). The upper molars of *S. elissae* have greatly reduced or absent styles and are more antero-posteriorly shortened in comparison to *Leptomeryx* (Figure 8). See Table 3 for comparative measurements of upper dentition.

Unlike *L. evansi,* the lower molars in *S. elissae* lack enamel crenulation and *Palaeomeryx*-folds but retain the *L. speciosus–L. evansi* morphology of the m3 entoconulid, as described by Korth and Diamond (2002) (Figure 2). Morphologically, the lower dentition of *S. elissae* is indistinguishable from *L. speciosus* (Table 1). *S. elissae* does follow an overall decreasing size trend, from *L. speciosus* to *L. evansi* to *S. elissae.* See Table 4 for comparative measurements of lower dentition. Lambe (1908) provides the only description of upper molars attributed to *L. speciosus*, which possess styles more akin to *L. evansi* than *S. elissae.* No specimen resembling either *L. speciosus* or *S. elissae* has been confidently identified from any early Orellan site, so if there is a direct relationship between *L. speciosus* and *S. elissae*, a record gap in the early Orellan makes it difficult to demonstrate such a transition.

Comparison to Pseudoparablastomeryx—Cranial features of *Santuccimeryx* most closely resemble the Arikareean leptomerycid *Pseudoparablastomeryx* (Tables 1 and 2). As in *Santuccimeryx, Pseudoparablastomeryx* demonstrates proportionally larger orbits. In *Pseudoparablastomeryx* (and in all leptomerycids known from cranial material), the postorbital bar is composed equally of the frontal and jugal. In *Santuccimeryx*, the jugal comprises the majority of the postorbital bar. Both genera demonstrate shallow and poorly developed postorbital constrictions,

TABLE 3 – Upper dentition measurements of *Santuccimeryx elissae*, *Leptomeryx evansi*, and *Pseudoparablastomeryx scotti*. All measurements in mm.

	BADL 64015 *Santuccimeryx elissae*	BADL 24824 *Santuccimeryx elissae*	SDSM 62381 *Leptomeryx evansi*	BADL 29030 *Leptomeryx* cf. *evansi*	YPM-PU 12962* *Leptomeryx evansi*	F:AM 95490 *Pseudoparablastomeryx scotti*
P2 – P4	N/A	N/A	20.0	17.6	17.5	14.9
P2 Length	6.2	N/A	N/A	N/A	6.0	5.1
P2 Width	3.7	N/A	N/A	N/A	4.0	4.0
P3 Length	5.3	5.9	6.9	6.3	9.0	5.1
P3 Width	4.8	4.8	5.8	4.9	5.0	5.3
P4 Length	N/A	4.2	5.8	4.8	5.0	4.5
P4 Width	N/A	5.8	6.4	5.9	5.7	5.9
M1 – M3	17.9	16.8	21.7	19.2	18.5	21.4
M1 Length	5.9	5.7	7.2	6.7	8.0	7.1
M1 Width	6.5	7.1	7.6	7.1	6.0	7.3
M2 Length	6.5	6.1	7.7	6.8	6.5	8.1
M2 Width	7.2	8.0	8.5	8.1	7.0	8.8
M3 Length	6.2	5.6	7.4	7.1	7.0	7.8
M3 Width	6.2	7.1	7.9	7.6	6.5	8.9

*measurements for YPM-PU from Scott (1940).

TABLE 4 – Lower dentition measurements of *Santuccimeryx elissae*, *Leptomeryx evansi*, and *Pseudoparablastomeryx scotti*. All measurements in mm.

	BADL 64015 *Santuccimeryx elissae*	Korth & Diamond (2002)* *"Leptomeryx" elissae*	SDSM 62381 *Leptomeryx evansi*	YPM-PU 12962** *Leptomeryx evansi*	Taylor & Webb (1976)*** *Pseudoparablastomeryx scotti*
p4 Length	6.0	N/A	6.3	6.0	6.1
p4 Width	3.1	N/A	3.6	3.0	3.1
m1 – m3	20.7	19.8	24.9	23.0	N/A
m1 Length	6.0	5.6	6.9	7.0	6.4
m1 Width	4.5	4.3	4.8	5.0	4.0
m2 Length	6.4	6.4	7.5	7.0	7.0
m2 Width	5.0	4.6	4.4	5.0	4.5
m3 Length	8.0	8.6	11.2	9.0	9.6
m3 Width	4.8	4.4	5.2	5.0	4.6

* measurements for *"Leptomeryx" elissae* are mean values from Korth and Diamond (2002).
** measurements for YPM-PU from Scott (1940).
*** measurements for *Pseudoparablastomeryx scotti* are mean values from Taylor and Webb (1976).

shortened crania, and shortened sagittal crests (Figures 6 and 7). The only skull of *Pseudoparablastomeryx* (F:AM 95490) has a 11.3 mm sagittal crest (Taylor and Webb 1976), comparable to the 10.3 mm sagittal crest of *Santuccimeryx* (BADL 64015) (Table 2). When comparing both genera to *Leptomeryx*, the distance between the frontal-parietal contact and anterior margin of the sagittal is increased as well as the angle between the temporal crests (64° in *Santuccimeryx*, 60° in *Pseudoparablastomeryx*) (Figures 6 and 7; Table 2). Reliable comparison of the basicrania is not possible because of the suboptimal preservation of the only known skull of *P. scotti* (F:AM 95490).

The upper premolars in *P. scotti* develop strong anterolingual cingula, giving them a submolariform appearance with strong anterior and posterior fossettes (Figure 8). The anterolingual cingulum and corresponding fossette are lacking in the P3 for *S. elissae*. The P3 postprotocrista in *S. elissae* is directed posterolaterally, creating a small basin, similar to what is observed in *Pseudoparablastomeryx* but not as strongly pronounced (Figure 8). The upper molars of *Santuccimeryx* have reduced or absent styles and are more antero-posteriorly shortened in comparison to *P. scotti* (Figure 8). See Table 3 for comparative measurements of upper dentition.

The lower dentition of *P. scotti* resembles that of *S. elissae* in that it has smooth enamel, a well-developed buccal pillar, and lacks the *Palaeomeryx*-fold. These characters were consistent with the holotype (F:AM 95490) and all observed, referable specimens of *P. scotti*. The m3 entoconulid in *P. scotti* is, however, greatly reduced when compared to *S. elissae* or *L. evansi*. The dp4 in *S. elissae* appears to be morphologically similar to the condition in *P. scotti*, particularly with cingulum and buccal pillars separating the anteroconid, protoconid, and hypoconid, as also noted by Taylor and Webb (1976). See Table 4 for comparative measurements of lower dentition.

MORPHOLOGICAL TRENDS

Relationships between Eocene and Oligocene leptomerycids continue to rely almost entirely on dental characters, specifically on the morphology of m3, which suggests two distinct lineages: the *L. yoderi–L. mammifer–L. exilis–L. obliquidens* lineage (Korth and Diamond 2002) and the *L. speciosus–L. evansi* lineage (Heaton and Emry 1996, Korth and Diamond 2002) (Figure 9). The *L. speciosus–L. evansi* lineage would also include the progression to *S. elissae*, as suggested by Korth and Diamond (2002) (Figure 9). We primarily agree with these assertions, given the characters provided by the material currently available. However, we believe that BADL 64015 does not bring further clarity to the *L. speciosus–L. evansi* relationship. If *S. elissae* stemmed from *L. evansi*, it would imply a new lineage that emerged in the early Oligocene and provides a link to later Oligocene and Miocene leptomerycids (Figure 9). However, because the dental characters in *S. elissae* are more akin to the condition in *L. speciosus* (Table 1), it is likely that *L. speciosus* gave rise to the two lineages of *L. evansi* and *S. elissae* at to around the Eocene–Oligocene transition (Figure 9). We recognize that once skulls of Eocene leptomerycids are discovered, additional cranial features may further complicate leptomerycid relationships. If a skull of *L. speciosus* were discovered and the cranial characters reflected a closer relationship to *S. elissae* in addition to the dental

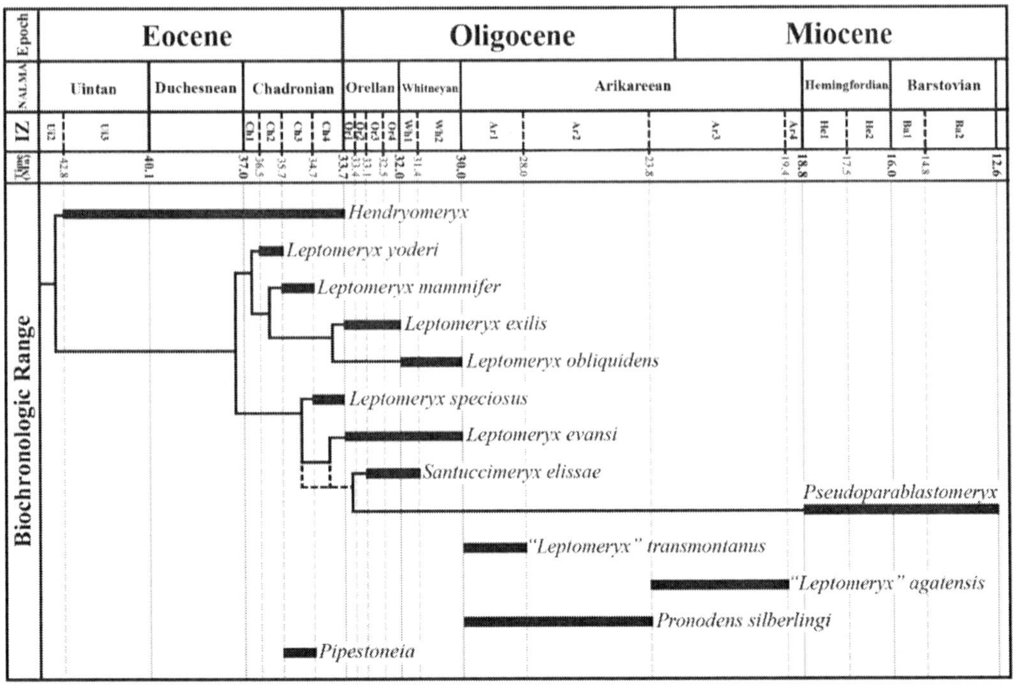

Figure 9. Biochronologic ranges of North American Leptomerycidae with hypothetical phylogenetic relationships inferred by previous authors (Heaton and Emry 1996, Korth and Diamond 2002). Hatched lines illustrate alternative branching points for the inferred *Santuccimeryx—Pseudoparablastomeryx* group. Other taxa of uncertain affinity are not positioned on branches.

characters, it could imply that the cranial anatomy of *L. evansi* is uniquely derived and not necessarily representative of other taxa currently referred to *Leptomeryx*.

Arikareean Leptomerycids: Leptomerycids identified from the Arikareean have been problematic in previous studies and are in need of careful review. Three taxa have been named from Arikareean faunae: *Leptomeryx transmontanus*, *Leptomeryx agatensis*, and *Pronodens silberlingi* (Figure 9 and Table 1). Douglass (1903) named *L. transmontanus* from a partial cranium with upper dentition from the Lower Cabbage Patch Beds of Montana (Ar1). The species was recognized by Frick (1937), but Webb (1998) suggested synonymizing *L. transmontanus* with *Hendryomeryx esulcatus*. Korth and Diamond (2002) suggest that *L. transmontanus* is synonymous with *Pronodens*, based on Rasmussen's (1977) reclassification in an unpublished dissertation. They argue that *Pronodens* likely stemmed from *L. evansi* based on a more prominent *Palaeomeryx*-fold but provide no specimens for reference. The type of *P. silberlingi* figured by Koerner (1940) and Taylor and Webb (1976) exhibits extremely worn molars that do not show features mentioned by Korth and Diamond (2002). There are no unambiguous characters to delineate *L. transmontanus* (Table 1). Therefore, on the basis of measurements alone, this species could just as easily be referred to *L. obliquidens* as to *Pronodens*.

L. agatensis is based on a partial maxilla from the *Syndyoceras* Quarry in Nebraska (Ar3) and has been synonymized with *L. obliquidens* due to morphologic similarities (Cook 1934, Taylor and Webb 1976, Webb 1998). However, Korth and Diamond (2002) again infer an equally close affinity to *Pronodens*. We maintain that, given the state of the *L. obliquidens* and *Pronodens* material, no confident synonymization is currently possible. In contrast to previous authors, Macdonald (1963, 1970) did not assign species-level designations to *Leptomeryx* specimens from the Wounded Knee Fauna (Ar1), the only other published Arikareean leptomerycid material, on account of a lack of unambiguous characters.

Aside from the procumbent incisors, no other characters link *Pronodens* to the Leptomerycidae (Table 1). The molars belonging to the holotype (YPM 13952) are extremely worn, rendering it impossible to ascertain any of the reliable characters utilized to distinguish the species of other leptomerycids. Since the molars and premolars do not provide defining characteristics, *Pronodens* is justified by the loss of p1, a shortened diastema between the canine and p2, and a more posterior position of the mental foramen—all characters found in the anterior portion of the dentary. This portion is the most likely feature of the dentary to be missing in leptomerycid specimens from the White River Group, again making it impossible to refer specimens to *Pronodens* or to synonymize *Pronodens* with other genera (Koerner 1940). Taylor and Webb (1976) attempted to refer one specimen (AMNH 13824, *Leptomeryx obliquidens*) to *Pronodens*, but this is tentative at best. *Pseudoparablastomeryx scotti* is the only other leptomerycid known to share the loss of p1, but it maintains a longer diastema between the canine and p2 despite being more brachycephalic. The dentaries of BADL 64015 are broken at the anterior side of p3, and all other referred *S. elissae* specimens are partial dentaries with cheek teeth. It is thus not possible to discern the presence of p1 or the diastema between the canines and remaining premolars without additional specimens containing these features.

Pronodens is also considered to be slightly larger than *Leptomeryx* (Webb 1998) based on molar measurements, but there is, of course, a limited sample size for this genus. Considering the shortening of the diastema and the premolars, *Pronodens* is likely brachycephalic compared to *Leptomeryx*. It is clear, however, that besides the type specimen, all specimens referred to as *Pronodens* are questionable and have no direct correlation to the type. It will be difficult to track leptomerycid diversity and evolutionary relationships from post-Whitneyan faunae until Arikareean taxa are clarified through additional material.

Convergence with the Tragulidae: *Santuccimeryx* exhibits a suite of characters that appear intermediate between *Leptomeryx* and *Pseudoparablastomeryx* (Table 1), while also demonstrating a morphologic trend of cranial features that appear to be convergent with the Eurasian and African Tragulidae. The postorbital bar is the most prominent example—in the North American leptomerycids, it is composed of equal positions of the frontal and jugal, but in *S. elissae*, the jugal comprises the majority. This trait has never before been documented in North American leptomerycids but has been used as a character used to delineate the Tragulidae from the rest of Traguloidea (Vislobokova 2001, Métais and Vislobokova 2007). The shallow and poorly-developed postorbital constriction in *S. elissae* (and in *Pseudoparablastomeryx*) is also observed in the extant tragulids, *Hyemoschus* and *Tragulus*. The basioccipital in *Santuccimeryx* is trapezoidal, again similar to the tragulid *Hyemoschus* and differing greatly from the subrectangular basioccipital in *Leptomeryx*. All of these characters—the jugal comprising most of the postorbital bar, the shortened sagittal crest, and the posteriorly expanded or trapezoidal basioccipital—have been used as synapomorphies to distinguish Tragulidae from other Traguloidea (Vislobokova 2001, Métais and Vislobokova 2007). Given that these cranial features are similar to what is documented in the evolution of the Tragulidae by previous authors (Vislobokova 2001, Métais and Vislobokova 2007), the character convergence of *Santuccimeryx* warrants future exploration.

CONCLUSIONS

The small, hornless, deer-like leptomerycids proliferated across the North American Great Plains from the late Eocene into the Miocene, leaving behind an abundant fossil record that has proven biostratigraphically valuable. Despite their significance and abundance, leptomerycid evolutionary relationships resist clarification due to both the scarcity of cranial material and the high degree of individual variation expressed in leptomerycid dentition. No non-dental cranial material has yet been reported from *Hendryomeryx*, *L. blacki*, *L. yoderi*, *L. mammifer*, *L. speciosus*, *L. exilis*, *Pronodens*, or *Pseudoparablastomeryx francescita*. Any studies into leptomerycid phylogeny up to this point have only had skulls from *L. evansi*, *L. obliquidens*, and *P. scotti* to reference and have thus been forced to rely on dental characters and temporal succession. The evolutionary relationships of the leptomerycids should thus be reevaluated if and when supplementary cranial material becomes available.

The addition of BADL 64015 to the known record of leptomerycid cranial material provides insights into evolutionary relationships that would not have been possible with only dental material, thereby proving that dental characters alone are problematic in delineating leptomerycid taxa and evolutionary relationships. Based solely on dental characters, BADL 64015 should undoubtedly be placed within *Leptomeryx*, as Korth and Diamond (2002) proposed, but observed cranial characters diverge vastly from *Leptomeryx* and instead suggest placement within *Pseudoparablastomeryx* and convergence with the Tragulidae. When we consider the intermediate nature of the anatomical traits exhibited by BADL 64015 in conjunction with its temporal intermediacy to *Leptomeryx* and *Pseudoparablastomeryx*, BADL 64015 not only validates Korth and Diamond's (2002) distinction of *Leptomeryx elissae* as a species but also elevates the taxon to a new genus: *Santuccimeryx*.

ACKNOWLEDGEMENTS

We thank, first and foremost, Tiffany Leone, the 2016 Geoscientists-in-the-Parks intern at Badlands National Park, who found and properly reported BADL 64015. Many thanks to Ellen Starck (Park Paleontologist), Wayne Thompson (Physical Science Technician [Paleontology]), Mary Carpenter (Fossil Preparation Lab Manager), and Alex Carrier (Museum Curator) at Badlands National Park for collection, preparation, access to specimens, and their continued support of the project. Additional gratitude is due to Mary Carpenter who flawlessly prepared BADL 64015 back in 2016. We are so grateful to Badlands Natural History Association for funding travel to Pittsburgh to present this research at the 2023 Geological Society of America Connects meeting. Thanks to Nate Fox, Darrin Pagnac, and Sam Wright at South Dakota School of Mines & Technology Museum of Geology for access to specimens and for being witness to our descent into leptomerycid madness. Thanks also to Judy Galkin and Meng Jin at the American Museum of Natural History for access to specimens, especially the *P. scotti* holotype. Last, but certainly not least, a huge thank you to Maya Feldberg-Bannatyne for her thoughtful edits, constant support, and willingness to listen to us ramble about tiny deer at any time of the day. Any opinions expressed in this study do not necessarily reflect the opinions of the National Park Service or the Department of the Interior.

LITERATURE CITED

Benton, R. C., D. O. Terry Jr., E. Evanoff, and G. McDonald. 2015. The White River Badlands: Geology and Paleontology. pp. 1-222. Indiana University Press, Bloomington, Indiana.

Bouvrain, G. and D. Geraads 1985. Un squelette complet de Bachitherium (Artiodactyla, Mammalia), de l'oligocène de Céreste (Alpes de Haute Provence). Remarques sur la systématique des ruminants primitifs. Comptes Rendus de L'Académie des Sciences, Paris, Sér. 2, 300(2):75–78.

Cook, H. J. 1934. New artiodactyls from the Oligocene and Lower Miocene of Nebraska. American Midland Naturalist 15:148-165.
Douglass, E. 1903. New vertebrates from the Montana Tertiary. Annals of the Carnegie Museum 2:145-199.
Evanoff, E. E., D. O. Terry Jr., R. C. Benton, and H. Minkler. 2010. Field guide to the geology of the White River Group in the North Unit of Badlands National Park. Pages 96-127 in Terry, M.P., E.F. Duke, and J.A. Tielke, editors. Geologic field trips in the Black Hills region, South Dakota: South Dakota School of Mines and Technology Bulletin 21.
Frick, C. 1937. Horned ruminants of North America. Bulletin of the American Museum of Natural History 69:1–669.
Gazin, C. L. 1955. A review of the upper Eocene Artiodactyla of North America. Smithsonian Miscellaneous Collections 128:1–96.
Gentry, A. W. and J. J. Hooker 1988. The Phylogeny of the Artiodactyla, in The Phylogeny and Classification of the Tetrapods, vol. 2: Mammals. Oxford: Clarendon, pp. 235–272.
Geraads, D., G. Bouvrain, and J. Sudre. 1987. Relations phylétiques de Bachitherium Filhol, ruminant de l'Oligocène d'Europe occidentale. Palaeovertebrata 17(2):43–73.
Heaton, T. H. and R. J. Emry. 1996. Leptomerycidae. pp. 581-608 in D. R. Prothero and R. J. Emry (eds.) The Terrestrial Eocene-Oligocene Transition North America. Cambridge University Press, New York, New York.
Janis, C. 1987. Grades and Clades in Hornless Ruminant Evolution: The Reality of the Gelocidae and the Taxonomic Position of Lophiomeryx and Bachitherium. Journal of Vertebrate Paleontology 7(2):200–216.
Koerner, H. E. 1940. The geology and vertebrate paleontology of the Fort Logan and Deep River Formations of Montana Part I. New Vertebrates. American Journal of Science 238(12):837-862.
Korth, W. W. 2014. Rodents (Mammalia) from the Whitneyan (middle Oligocene) Cedar Pass Fauna of South Dakota. Annals of the Carnegie Museum 82(4):373-398.
Korth, W. W. and M. E. Diamond. 2002. Review of Leptomeryx (Artiodactyla, Leptomerycidae) from the Orellan (Oligocene) of Nebraska. Annals of Carnegie Museum 71:107–129.
Lambe, L. M. 1908. The Vertebrata of the Cypress Hills, Saskatchewan. Contributions to Canadian Palaeontology, Geological Survey of Canada 3(4): 1-64.
Linnaeus, C. 1758. Systema Naturae per Regna Tria Naturae, Secundum Classes, Ordines, Genera, Species, cum Characteribus, Differentiis, Synonymis, Locis. Editio Decima 1:1-824.
Lull, R. S. 1922. Primitive Pecora in the Yale Museum. American Journal of Science 5:111-119.
Macdonald, J. R. 1963. The Miocene faunas from the Wounded Knee area of western South Dakota. American Museum of Natural History Bulletin 125:139–238.

Macdonald, J. R. 1970. Review of the Wounded Knee faunas of southwestern South Dakota. Bulletin of the Los Angeles County Museum of Natural History and Science 8:1-82.

Mathis, J. E. and B. J. MacFadden. 2010. Quantifying *Leptomeryx* (Mammalia, Artiodactyla) Enamel Surface Area Across the Eocene-Oligocene Transition in Nebraska. Palaios 25: 682-687.

Matthew, W. D. 1903. The fauna of the Titanotherium beds at Pipestone Springs, Montana. Bulletin of the American Museum of Natural History 19:197–226.

Métais, G. and I. A. Vislobokova. 2007. Basal Ruminants. pp. 177-188 in D. R. Prothero and S. E. Foss (eds.). The Evolution of Artiodactyls. Johns Hopkins University Press, Baltimore, Maryland.

Owen, R. 1848. Description of teeth and portions of jaws of two extinct Anthracotherioid quadrupeds (*Hyopotamus vectianus* and *Hyop. bovinus*) discovered by the Marchioness of Hastings in the Eocene deposits on the NW coast of the Isle of Wight: with an attempt to develope Cuvier's idea of the Classification of Pachyderms by the number of their toes. Quarterly Journal of the Geological Society of London 4:103-141

Prothero, D. R. and R. J. Emry. 2004. The Chadronian, Orellan and Whitneyan North American Land Mammal Ages. Pages 156-168 in M.O. Woodburne (ed.). Late Cretaceous and Cenozoic Mammals of North America: Geochronology and Biostratigraphy. University of California Press, Oakland, California.

Prothero, D. R. and K. E. Whittlesey. 1998. Magnetic stratigraphy and biostratigraphy of the Orellan and Whitneyan land-mammal "ages" in the White River Group. Geological Society of America Special Paper 325:39-61.

Prothero, D. R. and R. J. Emry. 1996. Camelidae. pp. 609-651 in D. R. Prothero and R. J. Emry (eds.) The Terrestrial Eocene-Oligocene Transition North America. Cambridge University Press, New York, New York.

Rasmussen, D. L. 1977. Geology and mammalian paleontology of the Oligocene-Miocene Cabbage Patch Formation, central-western Montana. Unpublished Ph.D. Dissertation, University of Kansas, Lawrence, Kansas.

Schultz, C. B. and C. H. Falkenbach. 1968. The Phylogeny of the Oreodonts. Part 1: Merycoidodontinae, Eporeodontinae, and Leptaucheniinae, Three subfamilies of oreodonts, with an appendix to the revision of the Merycoidodontidae, and Part 2. Summary and conclusions concerning the Merycoidodontidae. Bulletin of the American Museum of Natural History 139:1-498

Scott, W. B. 1899. The selenodont artiodactyls of the Uinta Eocene. Transactions of the Wagner Free Institute of Science, Philadelphia 6:1-121.

Scott, W. B. 1940. Artiodactyla. Transactions of the American Philosophical Society 28:363–746.

Simpson, G. G. 1945. The Principles of Classification and a Classification of Mammals. Bulletin of the American Museum of Natural History 85:1-350.

Spaulding, M., M. A. O'Leary, and J. Gatesy. 2009. Relationships of Cetacea (Artiodactyla) Among Mammals: Increased Taxon Sampling Alters Interpretations of Key Fossils and Character Evolution. PLoS ONE 4(9): e7062. https://doi.org/10.1371/journal.pone.0007062

Stevens, M. S., D. R. Prothero, C. Cleaveland, E. Welsh, K. Marriott, T. Htun, D. Balassa, S. M. Olson, K. I. Watmore, and D. Wheeler. 2023. Systematics of the Late Oligocene and Miocene Oreodonts (Merycoidodontidae: Artiodactyla). New Mexico Museum of Natural History and Science Bulletin 93:1-226.

Tabrum, A. R. and G. Métais. 2007. *Pipestoneia douglassi*, a new genus and species of selenodont artiodactyl from the Pipestone Springs area, Jefferson County, Montana. Bulletin of the Carnegie Museum of Natural History 39:83-96.

Taylor, B. E. and S. D. Webb. 1976. Miocene Leptomerycidae (Artiodactyla, Ruminantia) and their relationships. American Museum Novitates, 2596:1-22.

Vislobokova, I. A. 2001. Evolution and classification of Tragulina (Ruminantia, Artiodactyla). Paleontological Journal, 35:69–145.

Webb, S. D. 1998. Hornless ruminants. pp. 463–476 in C. M. Janis, K. M. Scott, and L. L. Jacobs (eds.) Evolution of Tertiary Mammals of North America, Volume 1: Terrestrial Carnivores, Ungulates, and Ungulatelike Mammals. Cambridge University Press, Cambridge, U.K

Webb, S. D. and B. E. Taylor. 1980. The phylogeny of hornless ruminants and a description of the cranium of *Archaeomeryx*. Bulletin of the American Museum of Natural History 167:117–158.

Wilson, J. A. 1974. Early Tertiary Vertebrate Faunas, Vieja Group and Buck Hill Group, Trans-Pecos Texas: Protoceratidae, Camelidae, Hypertragulidae. Bulletin of the Texas Memorial Museum-34.

Wortman, J. L. 1898. The extinct Camelidae of North America and some associated forms. Bulletin of the American Museum of Natural History 10(7):93-142.

Zanazzi, A., M. J. Kohn, and D. O. Terry. 2009. Biostratigraphy and Paleoclimatology of the Eocene-Oligocene boundary section at Toadstool Park, northwestern Nebraska, USA. The Geological Society of America Special Paper 452:197-214.

Zittel, K. A. 1893. Handbuch der Paläeontologie. Abteilung I. Paläeozoologie. Band IV, Vertebrata (Mammalia). R. Oldenbourg, Munich, Germany, 799 pp.

Abstracts of Research Papers and Posters

submitted for presentation at the

108th Annual Meeting

of the

South Dakota Academy of Science

BIOSYNTHESIS AND APPLICATION OF PLANT NATURAL PRODUCTS

Daniel K. Owens
South Dakota School of Mines and Technology
Rapid City, SD, USA
daniel.owens@sdsmt.edu

ABSTRACT

Examination of the flavonoid biosynthetic pathway continues to provide information that is being used to improve the quality of agricultural, medicinal, and nutritional products. Furthermore, this analysis provides foundational insights into the genetics, enzymology, and metabolic organization of natural product biosynthesis. The focus and unifying theme of our research is the study of natural products that are important for the survival and development of plants and that can be utilized to the advantage of society. One important aspect in achieving these far-reaching goals is to unravel and characterize the organization of pathways leading to the accumulation of specific natural products *in planta*. To this end, we are elucidating metabolon formation of the flavonoid biosynthetic pathway in *Citrus sinensis* (Orange). This work includes not only enzymes involved in production of the C6-C3-C6 flavonoid core structures, but also derivatizing enzymes, such as glycosyltransferases. Derivatizing enzymes are of particular interest as they ultimately lead to the biosynthesis of the myriad compounds which accumulate *in planta*. In related work, we have identified plants with allelopathic activity from a Hawaiian ecosystem using a phylogenetic approach subsequently verified by direct bioassay. Of particular note, an allelopathic compound from *Psidium cattleianum* (Strawberry Guava) that contributes to it invasive strategy has been identified, purified, and characterized.

DO WE NEED A SALINE BOLUS?— THE NECESSARY REFINEMENT OF PEDIATRIC PROPOFOL SEDATION

Andrew Guymon[1], Madigan Moore[1], Laura Nelson[1], and Mir Ali[2]
[1]University of South Dakota Sanford School of Medicine
[2]Sanford Children's Hospital
Sioux Falls, SD 57105
Corresponding author email: andrew.guymon@coyotes.usd.edu

ABSTRACT

Background: US healthcare costs have risen dramatically, encompassing almost 20% of the nation's GDP, and costs of sedations are no exception. Propofol sedation is used ubiquitously for pediatric MRIs when intensivists must ensure minimal patient movement for high-quality pictures. Normally, patients receive a 20 ml/kg saline bolus in hopes of maintaining blood pressure throughout propofol sedation. The goal of this study was to determine if a 10 ml/kg saline bolus was as equally efficacious as a 20 ml/kg saline bolus at completing pediatric propofol sedations and maintaining MAP in aims of reducing healthcare costs.

Methods: The study had two phases. The first phase was a six-month review of saline bolus dosing using the standard 20 ml/kg bolus. The second phase introduced a goal saline bolus dose of 10 ml/kg. A successful sedation was determined by completing the MRI without any complications or signs of cardiovascular distress.

Results: A total of 114 patients ranging from ages 18 months to 16 years were recruited. Percentage of successful sedations with a 20 ml/kg bolus and 10 ml/kg bolus was 94% and 100%, respectively. The average total saline bolus was reduced by 49% from phase 1 to phase 2. Importantly, the average MAP throughout sedation (20 ml/kg: 57±2 mmHg, 10 ml/kg: 58±2 mmHg, $p = 0.425$), minimal MAP (20 ml/kg: 50±2 mmHg, 10 ml/kg: 52±1 mmHg, $p = 0.125$) and end-sedation MAP (20 ml/kg: 58±1 mmHg, 10 ml/kg: 57±1 mmHg, $p = 0.610$) were equivocal between the two groups.

Conclusion: We conclude that a protocol with a baseline saline bolus dose of 10 ml/kg would preserve blood pressure, allow for successful sedation, but prevent unnecessary excess dosing.

BMP9/ALK1 SIGNALING IN HUMAN HIGHLY PROLIFERATIVE CELLS

Oleksandra Rachynska, Calvin Vary, and Michayla Moore
Dakota State University
Madison, SD 57042
Corresponding author email: Calvin.vary@dsu.edu

ABSTRACT

The leading cause of cardiovascular disease mortality is myocardial infarction (MI). MI leads to the sufficient loss of cardiac cells and, eventually, granulation tissue formation. The outcomes of the granulation tissue appearance are the limited contractive abilities of the heart muscle and insufficient blood supply. Angiogenesis, new blood vessel formation, can support the restoration of blood flow in the infarcted heart area. Targeting the angiogenic mechanisms is crucial for the repair of MI. We have found that human cardiac-derived highly proliferative cells (hHiPC, Ryzhov et al.) express the BMP9/10 receptor, ALK1, which can be targeted activated by BMP9 treatment. We studied whether BMP9 treatment that targets angiogenesis genes in hHiPC cells can promote the pro-angiogenic and pro-reparative secretome, which repairs injured cardiac tissue. To test the role of the ALK1 receptor in this process, we examined whether the ALK1 inhibitor LDN193189 (Horbelt et al.) would inhibit BMP9/ALK1 induction of putative hHiPC proteins. Specifically, we examined the response of mRNAs encoding putative BMP9 target proteins CD105, SOST, IGFBP3, as well as ALK1. Using RNA isolation following BMP/LDN treatment, cDNA synthesis, and RealTime qRT-PCR, we found that LDN193189 treatment reduced the response of BMP9-dependent induction of CD105, SOST, and IGFBP3 mRNA, and not ALK1 mRNA. These results confirmed previous findings (Moore et al., unpublished) and further support the hypothesis that hHiPC ALK1 is essential for the induction of these mRNAs, and that LDN193189 is a useful reagent for the inhibition of BMP receptor signaling in hHiPC. To gain a broad view of LDN193189-dependent effects on hHiPC BMP signaling via ALK1, we performed proteomic analysis by liquid chromatography/mass spectrometry of BMP9- and LDN-treated hHiPC secreted proteins. Still in progress, initial data detected a previously hHiPC BMP9 target, ISLR/Meflin, and as well as several novel potential BMP9 hHiPC secreted proteins, including Inhibin-beta (Jaatinen et al.) and CCN2. Inhibins can regulate BMP9 signaling (Huang et al.). These results may shed light on the role of ALK1 signaling to the overall capacity of hHiPC to contribute to cardiac repair following MI.

DIRECT AIR CAPTURE VIA METAL-ORGANIC SUPERCONTAINERS

Hanying Li and Zhenqiang Wang*
Department of Chemistry
University of South Dakota
Vermillion, SD 57069
*Corresponding author email: hanying.li@coyotes.usd.edu

ABSTRACT

Society's continued demand for fossil fuel consumption has led to a drastic increase in atmospheric carbon dioxide levels, causing rising global air temperatures. To combat global warming, post-combustion carbon capture techniques have been developed to mitigate CO_2 emissions at point sources such as fossil-fuel fired power plants, though these methods do not address CO_2 already emitted to air. Direct removal of CO_2 from ambient air, known as direct air capture (DAC), represents a net negative carbon emissions technology and offers an opportunity to reverse the current trend of global warming. However, developing viable DAC technologies remains a daunting task due to the ultra-dilute concentration of atmospheric CO_2 and the prohibitively high costs associated with the "covalent" acid-base chemistry underlining the current DAC methods.

We recently discovered that a family of cage-like molecules known as metal-organic supercontainers (MOSCs) show an encouraging capacity to capture CO_2 directly from ambient air. MOSCs are synthetic host molecules featuring both *endo-* and *exo-*cavities suitable for selectively binding small molecules and ions. One particular subset of MOSCs, obtained from the assembly of divalent and trivalent metal ions (e.g., Zn^{2+}) bridged by suitable organic linkers (e.g., pyridines), show a promising ability to directly capture CO_2 from ambient air. Nuclear magnetic resonance (NMR) studies suggest that the MOSCs dissolved in a dimethyl sulfoxide (DMSO) solution binds ~6 equivalents of CO_2 from the atmosphere. Preliminary mechanistic studies indicate that non-covalent interactions, such as hydrogen bonding, play a key role contributing to the DAC. This MOSC-based, "non-covalent" DAC approach thus has the potential to overcome some of the limitations of the current methods. Future studies will focus on determining the exact carbon capture mechanism and improving the DAC efficiency of the MOSCs.

INCORPORATING INSTRUMENTATION IN LARGE ENROLLMENT GENERAL CHEMISTRY LABORATORIES

R. C. Engstrom[1] and M. D. Koppang[2]
[1]Department of Chemistry
University of Montana
Missoula, MT
[2]Department of Chemistry
University of South Dakota
Vermillion, SD 57069
Corresponding author email: mkoppang@usd.edu

ABSTRACT

We introduced departmental instrumentation into the University of South Dakota General Chemistry Laboratory which consisted of multiple lab sections of 56 students per section. Two unique labs were developed and implemented including a lab on chemical microscopy and a lab on isotopes. These labs were also used in our Honors Chemistry lab. The first lab on chemical microscopy utilized polarized microscopes equipped with cameras and monitors. These microscopes were obtained by funds to RCE from a National Science Foundation grant. Crystals were grown from saturated aqueous solutions placed on microscope slides and viewed during early stages of crystal growth. Microscopy also focused on sublimed crystals of iodoform and deposition of silver dendrimers on copper wire. Various aspects of microscopy and light were discussed including plane polarized light and isotropic versus anisotropic with respect to rotation of plane polarized light by crystals. The second lab focused on isotopes. Using IR spectroscopy, students observed, measured and compared to Hook's Law the C-H vs. C-D bond of chloroform versus deuterated chloroform . Hook's Law was also demonstrated with different masses on a spring and measurement of their different oscillation frequencies. Finally, the isotopes of halogens using chlorobenzene and bromobenzene were detected and measured with mass spectrometry.

INTERACTION OF ENVIRONMENTAL FACTORS ON MACROINVERTEBRATE BODY SIZE

Aria Smith, Vojsava Gjoni, Jeff Wesner
Biology Department
University of South Dakota
Vermillion, SD 57069
Corresponding author: aria.smith@coyotes.usd.edu

ABSTRACT

Across ecosystems, body size changes seem to be regulated by temperature. However, body size is likely to be regulated by organisms in response to resource (food) availability as well. These changes can be measured across communities using size spectra, which describes the relative abundance of large to small individuals in an ecosystem. To understand how size spectra may change with different temperatures and potentially be offset by nutrient levels, we conducted a mesocosm experiment in the summer of 2022. Mesocosms were large, outdoor controlled tanks that simulate natural stream ecosystems; macroinvertebrate species naturally colonize them (Experimental Aquatic Research Site at USD). The experiment consisted of twenty-four tanks using a full-factorial design with two levels of temperature (unheated vs. heated), two levels of nutrients (ambient vs. enriched), and their interaction (n = 6 replicates each). Samples of macroinvertebrates from each tank were measured for body length using the program ImageJ. From the heated tanks, the relative abundance of larger organisms was much higher than expected from theory, but the increase depended largely on whether nutrients were added or not. The relative abundance of large individuals increased with temperature, but only in the tanks without added nutrients. With added nutrients, the relative abundance of large individuals decreased. These findings complicate predictions of food web change in response to nutrients and temperature, which have predicted reductions in large individuals with temperature unless there is a corresponding increase in resource supply, the opposite of our findings.

DUST EMISSION POTENTIAL OF SOUTH DAKOTA SOILS: INSIGHT INTO THE 1930S DUST BOWL DROUGHT

Mark Sweeney[1*], Steve Forman[2], and Taylor Weeden[3]
[1]Dept. of Sustainability & Environment
University of South Dakota
Vermillion, SD 57069
[2]Dept. of Geosciences
Baylor University
Waco, TX 76798
[3]Umpqua Research Company,
Myrtle Creek, OR 97457
*Corresponding author email: mark.sweeney@usd.edu

ABSTRACT

The 1930s Dust Bowl Drought (DBD) was a climate crisis across the U.S. Great Plains. Attention of this 20th century megadrought has focused principally on the Southern Great Plains, the classic core of the Dust Bowl Drought. However, equally severe drought conditions, dust storms, and economic devastation occurred in the Northern Great Plains. In South Dakota alone, about 732 dust storms, severe wind erosion, and multi-year drought conditions were documented from 1930-1940, with heightened dust storm activity between 1933-1934. We used the Portable in situ Wind Erosion Lab (PI-SWERL) on soils in Dust Bowl-impacted counties of South Dakota to assess processes responsible for generating the "black blizzards" typical of the 1930s. High PM_{10} dust flux measured by PI-SWERL underscore the susceptibility of these soils to wind erosion with windspeeds of > 20 m/s, typical of dust storms during the DBD. Dust emissions are typically driven by saltation bombardment of sand grains entrained by the wind. Localized areas of dune sand such as in Tripp, Sanborn, and Brown counties, likely drove dust storm generation during windy days, but in Beadle County, one of the hardest hit counties in South Dakota, only 10% of the soils are sandy. Many soils in South Dakota are dominated by silt- and clay-rich shales and glacial till that produce sand-sized aggregates composed of silt and clay during wetting-drying and freeze-thaw cycles. Aggregates mobilized by the wind produce dust as they disintegrate upon impact with the soil surface or other grains. Beadle County soils contained ample sand-sized aggregates, notable in today's soils and documented by photographs from the 1930s. These results are important in identifying dust sources and processes that cause dust storms, especially during past droughts and forecasted aridification for the Great Plains in the 21st century.

CYTOTOXICITY AND DETERMINATION OF THE MINIMUM BACTERICIDAL CONCENTRATION (MBC) FOR *COLEUS SCUTELLARIOIDES*

Lexa Burtzlaff and Kathleen A. Gibson
Mount Marty University,
Yankton, South Dakota USA
*Corresponding author email: kathleen.gibson@mountmarty.edu

ABSTRACT

Antimicrobial resistance has become a rising issue in medicine, and plants have started to be seen as a possible solution. The purpose of this study was to determine whether *Coleus scutellarioides* was able to inhibit certain bacteria in a Kirby Bauer assay and a minimum inhibitory concentration (MIC) assay. A cytotoxicity assay was performed to determine if the plant was damaging to cells, and if so, at what concentration. The results indicated that *C. scutellarioides* inhibited *Staphylococcus aureus* and *Bacillus subtilis* with a tentative MIC of 125 mg/ml. The cytotoxicity assay determined that *C. scutellarioides* was not cytotoxic to eukaryotic cells.

DETERMINATION OF MINIMUM INHIBITORY CONCENTRATION (MIC) AND MINIMUM BACTERICIDAL CONCENTRATION (MBC) OF *SALVIA ROSMARINUS* USING A 96 WELL PLATE ASSAY

Elisabeth McGill and Kathleen A. Gibson
Mount Marty University,
Yankton, South Dakota USA
*Corresponding author email: kathleen.gibson@mountmarty.edu

ABSTRACT

This research investigated the antimicrobial properties of secondary metabolites in *Salvia rosmarinus* (rosemary) through screening analysis and bioassays. Extracts of *S. rosimarinus* were prepared using two solvents: boiling water and 75% ethanol. These extracts were screened for bacterial inhibition using Kirby-Bauer disk diffusion assay and against *Artemia salina* for cytotoxicity. Minimum inhibitory concentration (MIC) and minimum bactericidal concentration (MBC) were determined using a 96 well plate assay. Ethanol extracts showed the greatest inhibition of *Bacillus subtilis* and *Staphylococcus aureus* whereas the water extracts did not inhibit *B. subtilis* and partially inhibited *S. aureus*. Water extracts and lower concentrations of the ethanol extract were not cytotoxic. At 1000 μg/ml, *S. rosimarinus* was partially cytotoxic with at 38% survival rate compared to Gentamicin's survival rate of 88%. Preliminary results also indicate an MIC of 3.9 mg/ml and MBC of 7.8 mg/ml for *B. subtilis*.

WHICH IS BEST: ONLINE, FACE TO FACE OR HYBRID EDUCATION?

Pankaj Mehrotra*
University of the People
Pasadena, CA 91101
*Corresponding author email: mehrotra85pankaj@gmail.com

ABSTRACT

Learning Spaces are important for students to interact with peers and professors. Learning spaces are the zones interaction between students-students and professors-students. Both pre and post COVID-19 shift in paradigm from instructional based teaching to active learning has evolved at a rapid pace due to development of new technologies, addition of new content and shift in composition of student population from homogeneous to heterogeneous. Enhanced learning outcomes may be accomplished through new ways of teaching, active learning activities and assessment have been implemented to educate both ESL and non-ESL students. I will be presenting my original work of teaching, learning and assessment in asynchronous, synchronous and in-person courses I have taught in the US, other parts of the world and also virtually. Selection of teaching methods & technology such as chalk & talk, touch screen, projector, board varies with number and type of student population. Thoughtful implementation of active learning activities aids in retention and then enhancement of student learning. Challenging students with constructive and selective response question with high-order complexity improves growth mindset and critical thinking. The focus of the presentation will also be on how to form better student groups in STEM courses by changing the approach and structure of the student group. Overall, both online and face to face education are important to facilitate student learning and also for the development of educators.

LOSSY TRANSMISSION LINES AS A TEACHING TOOL FOR THE FINITE-DIFFERENCE TIME-DOMAIN (FDTD) METHOD

Thomas P. Montoya*
Department of Electrical Engineering & Computer Science
South Dakota School of Mines & Technology
Rapid City, SD 57701-3995
*Corresponding author e-mail: Thomas.Montoya@sdsmt.edu

ABSTRACT

This presentation will show results on using lossy transmission lines as a teaching tool for the Finite-Difference Time-Domain (FDTD) method. Commonly used by the electromagnetics community, the FDTD method is a numerical method for directly computing time-domain solutions to Maxwell's equations, i.e., Ampere's and Faraday's Laws. The FDTD method takes these spatial and time partial differential equations and expresses them in a discretized form, allowing for the direct solution of the electric and magnetic field components. The field quantities are interleaved in both space (unit cells) and time in a fashion that allows for successive updating (via update equations) with a 'time-stepping' or 'time-marching' algorithm. The FDTD method is computationally intensive in terms of memory and processing power. For example, three dimensional (3D) problems can have six field components and three material coefficients per unit cell. Therefore, an $N \times N \times N$ FDTD mesh requires ~$9N^3$ items be stored in memory and ~$6N^3$ items updated at each time step. In contrast, using lossy transmission line theory to introduce the FDTD method builds on the well-known Telegrapher's equations. The two Telegrapher's equations are one dimensional spatial and time partial differential equations involving two variables (voltage and current) and four parameters. Therefore, an N-long FDTD mesh for a lossy transmission line will require ~$6N$ items to be stored in memory and ~$2N$ items updated at each time step. This provides a means for introducing students to the FDTD method that is much less demanding from both a theoretical as well as a computational standpoint.

FIELD-BASED EVALUATION OF ROOT SEGMENT RESPROUTED YELLOW-FLOWERED ALFALFA STAND: PERSISTENCE AND BIOMASS

Abdulallah M. Alomair[1], Arvid Boe[2], and Lan Xu[1]*
[1]Department of Natural Resource Management
[2]Department of Agronomy Horticulture & Plant Science
South Dakota State University
Brookings, SD 57006
*Corresponding Author: Lan.Xu@sdstate.edu

ABSTRACT

Alfalfa is a crucial forage crop for livestock producers and dairy enterprises in the USA and worldwide due to its high nutritional value for animals. However, alfalfa planting is limited by drought, winter hardiness and grazing in semiarid and arid regions. This study aims to evaluate the phenomenon of yellow-flowered alfalfa (*Medicago falcata*) developing adventitious shoots on root segments under field conditions. This trait is of interest because it has the potential to enhance alfalfa persistence and productivity in semiarid and arid working lands. The research was conducted at South Dakota State University Field Station in Aurora and involved 14 yellow-flowered alfalfa populations from the National Plant Germplasm Core Collection (NPGCC). Eight-nine transplants used in the study were derived from root segments that had resprouted in a greenhouse five years earlier, with each spaced 3 feet by 3 feet. These root segments were originally 6 cm long and were taken from 1 cm and 7 cm below the cotyledonary node of one-year-old plants. After two years of transplantation, we measured various parameters including: percentage of survival, flower and seed pod production, flower color, crown area, growth habit, canopy volume, and biomass in the field. We found that 98% of the transplanted plants survived. All populations produced flowers and seeds. Two populations, PI 502441 and PI 538984 produced the highest dry biomass (240 g/m^2, equivalent to 2.64 tons/ha) with larger crown area and canopy volume. The results suggest that the plants generated from roots can self-sustain under field conditions. Furthermore, the study underscores the importance of field testing to identify additional populations within the NPGCC that contribute persistent forage to the landscape by maintaining and regenerating stands with strong root sprouting capacity. This information can be valuable for farmers, ranchers, and land managers.

INSTRUCTIONS FOR AUTHORS

Editorial Policy: The South Dakota Academy of Science is a forum for the promotion of scientific teaching, research, and service in the state of South Dakota. The South Dakota Academy of Science meets each spring for the purpose of annual business, awards, and the interchange of ideas, information, and results from scientists and students of science. The minutes of the annual meeting and the abstracts of presentations are published annually in the Proceedings of the South Dakota Academy of Science. In many cases, authors choose to submit a full manuscript of their presentation, which is published in lieu of the abstract. **Deadlines for submission of full manuscripts are July 15** each year; to submit manuscripts, please adhere to the Submission Procedures listed below. Authors have the option of purchasing additional reprints; information on page and reprint charges will accompany the page proofs sent to the corresponding author. The following guidelines provide instructions for authors submitting manuscripts to the Proceedings. Be aware that these guidelines may vary from year to year, so please try to format your manuscript so that it meets the conventions described below.

Authorship Policy: The following is the policy on authorship of abstracts and manuscripts to be published in the Proceedings:

1) All papers published in the Proceedings of the South Dakota Academy of Science will include all authors that were shown on originally submitted abstracts or manuscripts.

2) Any request to change authors after a submission has been made to the Proceedings must be accompanied by a brief letter of explanation sent to the editor that includes the original signature of all authors (including authors removed from or added to the authorship list).

3) The deadline for requesting changes in authorship is 30 August of the year of submission.

Manuscript Submission Procedure: Manuscripts submitted for publication in Proceedings must be accompanied by **two** technical reviews. Authors are required to solicit reviewers that have expertise in the topic area of the manuscript. Co-authors of manuscripts are not considered reviewers. Each reviewer must provide their name, title, address, and phone number along with a brief review of the manuscript; to facilitate this process, the reviewers should use the Peer Review Form that can be downloaded from the Academy's web site (http://sdstate.edu/sdas/). In turn, authors must provide a brief reconciliation letter describing how reviewer comments were addressed along with a final, revised copy of the manuscript. The submission package must include the following: 1) two technical reviews (Peer Review Forms), 2) written reconciliation from authors, and 3) revised manuscript. All file names should begin with the corresponding author's last name. Failure to adhere to the submission procedure will result in manuscripts being returned to authors.

Submission of abstracts for publication in the Proceedings **do not need** to be peer-reviewed.

Abstracts or manuscripts may be submitted either electronically or by mail. Please do not submit abstracts or manuscripts both electronically and by mail.

Electronic submission:
Electronic submissions should be forwarded to **Robert Tatina at rotatina@dwu.edu**. Manuscripts should be saved as **MS Word files** (not PDFs) that include tables and/or figures. Reviewer comments and the author(s) reconciliation statement must be submitted as separate files (MS Word) along with the manuscript. Manuscript illustration files that have the extension .BMP, .GIF, .EPS, .JPG, .TIF, or .PCX are acceptable. IMPORTANT: **Begin all file names with the corresponding author's last name** (e.g., Chipps_ms.doc, Chipps_review1.doc, Chipps_review2.doc, Chipps_reconcil.doc).

Mail submission:
Manuscripts may be submitted by mail by including one hardcopy and one electronic copy (3.5" disk or CD) of each manuscript. Hardcopies of reviewer's comments and reconciliation statement must also be included. Each diskette or CD should be identified with authors' names and by the format used. Submit all information in MS Word. Illustration files that have the extension .BMP, .GIF, .EPS, .JPG, .TIF, or .PCX are acceptable. **Begin all file names with the corresponding author's last name** (e.g., Chipps_ms.doc).

Mail manuscript submissions to:

Robert Tatina, Proceedings Editor
308 Orchard Blvd.
Sawyer, MI 49125

Phone: 269-426-3744
E-mail: rotatina@dwu.edu.

Membership. The author or one co-author of each abstract or paper must be a current member of the South Dakota Academy of Science.

Web Site Address. http://www.sdaos.org

Format Conventions: All manuscripts and abstracts should be single-spaced throughout including tables and the literature cited section. All pages should be numbered serially in the upper right-hand corner. Use a 1-inch margin all around but do not justify the right margin or hyphenate words on the right margin (i.e., use word wrap). Avoid footnotes in both articles and abstracts. Use metric measurements unless English units are clearly more appropriate, in which case metric equivalents must appear in parentheses. Give scientific names for all species the first time the species' common name is used as follows: Culver's root (*Veronicastrum virginicum*). Write out numbers under 10, except in a series with numbers >10 or with measurements or percentages. Statistical terms and other measures should conform to the Council of Science Editors Scientific Style and Format (e.g., t = 0.311, df = 33, P = 0.001, note italics, capital P and spaces before and after =).

Manuscript Components:

TITLE PAGE
All manuscripts should include a title page that includes the author name(s), affiliation(s), current mailing address(es), and Corresponding author email. Here are the components and style of the title page:

Microsoft Word
Times New Roman, Font Size = 12
Margin: Left = 0, Right = 6, Paragraph Indent = 4
Title: All Caps, Bold, Centered
Authors: See example:

Alfred Jones[1*], William Smith[1], and Lawrence Porter[2]
[1]Biology Department
Cognation University
South Chicago, IN 44230
[2]Chemistry Department
Intuition College
Newell, SD 57330
*Corresponding author email: alfred.jones@cogun.edu

ABSTRACT

Include a brief but informative abstract. The abstract should be a single paragraph of 200-300 words that concisely summarizes the results and conclusions of the study. The abstract should be followed by a short list of keywords (< 5). For authors submitting only an abstract, you may include a few literature citations, but please use sparingly, and enclose them in parentheses within the text of the abstract.

Keywords

(Up five keywords)

INTRODUCTION

The introduction should include sufficient detail to establish the purpose and importance of the work being summarized. It should include pertinent literature related to the study question and end with a statement of the objectives/hypotheses being examined.

METHODS

The methods should include sufficient information for the reader to evaluate the data or repeat the study. It is better to be inclusive (i.e., overly explicit) when describing the methods used, but keep in mind that clarity and conciseness are important.

RESULTS

Results follow the methods section and should include a concise summary of your findings. Be careful not to 'discuss' implications of results; reserve these points for the Discussion section. Results that are presented in tables or figures should not be exhaustively discussed in the text. When statistics are presented they should be explicit (e.g. $t = 34.07$, df = 48, $P = 0.003$).

DISCUSSION

The common trait among good scientific papers is a well-written discussion section. The discussion section is where you synthesize what has been learned from the study and provide direction or generalities for future studies. The discussion section should not be merely a summary of relevant literature linked to restatements of your results. Rather, it should be an informative essay that advances what is known about the subject area. Be careful, however, not to extend the discussion beyond what can be supported by the data reported in the results.

LITERATURE CITED

All references cited in the text should be listed at the end of the manuscript under the Literature Cited heading. References should be in alphabetical order by the last name of the author(s). If different works by the same author(s) are referenced, references should be in chronological order. Authors cited in the text of abstracts or papers should be indicated in parentheses by last name and year of publication (e.g., Wade 1967; Bremer 1977). When a citation has more than two authors, the last name of the first author followed by et al. should be used (e.g., Seabloom et al. 1978). Where two or more papers by the same author(s) have appeared in one year, the style should be (Stewart 1967a, 1967b). Citations included in abstract only submissions should be complete and included within the text of the abstract.

Examples of Literature Cited

Anderson, R.J., and R.W. Schrier. 2000. Acute renal failure. Pages 1149-1160 *in* Braunwald, E., K.J. Isselbacher and R.G. Petersdorf, editors. Harrison's principles of internal medicine. 15th ed. McGraw-Hill, New York, NY.
Bremer, P.E. 1977. Pelican kill. Loon 49:240-241.
Gipson, P.S., I.K. Gipson, and J.A. Sealander. 1975. Reproductive biology of the bobcat (*Lynx rufus*).Journal of Mammalogy 56:605-612.
Green, J.C., and R.B. Janssen. 1975. Minnesota birds: where, when, and how many. University of Minnesota Press, Minneapolis, MN.
Midwest Regional Climate Center. 2002. Historical climate summaries. Available at <http://www.hprcc.unl.edu/> [Cited 12 May 2012]
Stewart, A.V., A. Joachimiak, and N. Ellison. 2009. Genomic and geographic origins of timothy (*Phleum* sp.) based upon ITS and chloroplast sequences. Pages71-81*in* T. Yamada and G. Spangenberg, editors. Molecular breeding of forage and turf. Springer, New York, NY.
Svihovec, L.K. 1967. A comparison study of the ecological distribution of small mammals in southwestern North Dakota. Thesis. University of North Dakota, Grand Forks, ND.

(For Internet citations, include as much information as possible so that readers could access the material cited. Also include "Available at" and then include the <URL>. Finally, add in square brackets the date you accessed the URL [Cited 6 June 2010].

Tables, Figures, Illustrations and Photographs. Tables should be double spaced throughout, without vertical lines, and with page numbers in the upper right-hand corner. If tables are more than one page in length, put "Table x. continued" on all subsequent pages. Each table should begin with a Table number (e.g., Table 1, Table 2, etc…) followed by a descriptive caption which is separated from the table headings and data by a horizontal line. Tables must be referenced in the text.

Figures should be done with black ink on firm quality paper or if computer generated, they should be of camera-ready quality (laser printer) with sharp lines suitable for reproduction. Photographs should be 5x7-inch, sharp, black and white glossy prints. When appropriate, a legend of symbols, etc., scale and a cardinal direction (e.g. North) should be indicated on figures of maps. On the back side, identify each figure or photograph lightly in pencil with the figure's number, author(s) name(s), and an indication of the top of the figure. Figure captions should be numbered (e.g., Figure x.) and typed on a separate sheet of paper at the end of the manuscript. Letters, symbols, and legends should be large and clear enough on all figures and photographs to be legible after copy reduction. Figures must be referenced in the text.

A note on deadlines: If you cannot meet the July 15th deadline for full papers, you must send the editor a note requesting an extension. You must also request a new due date because extensions are not automatically granted.

South Dakota Academy of Science
308 Orchard Blvd.
Sawyer MI 49125

ISBN 979-8-3313-0003-6

CURRAN ASSOCIATES INC.
proceedings
.com

Eighth International Conference on eHealth, Telemedicine, and Social Medicine (eTELEMED 2016)

Held at DigitalWorld 2016

Venice, Italy
24 - 28 April 2016

IARIA